Construction Technology
Analysis and Choice

Tony Bryan
University of the West of England

Blackwell
Publishing

Blackwell Publishing editorial offices:
Blackwell Publishing Ltd, 9600 Garsington Road, Oxford OX4 2DQ, UK
  Tel: +44 (0)1865 776868
Blackwell Publishing Inc., 350 Main Street, Malden, MA 02148-5020, USA
  Tel: +1 781 388 8250
Blackwell Publishing Asia Pty Ltd, 550 Swanston Street, Carlton, Victoria 3053, Australia
  Tel: +61 (0)3 8359 1011

First published 2005 by Blackwell Publishing Ltd

ISBN-10: 1-4051-2295-1
ISBN-13: 978-1-4051-2295-5

Library of Congress Cataloging-in-Publication Data
Bryan, Tony, 1947–
    Construction technology : analysis and choice / Tony Bryan.—1st ed.
      p.   cm.
    Includes index.
    ISBN-10: 1-4051-2295-1 (pbk. : alk. paper)
    ISBN-13: 978-1-4051-2295-5 (pbk. : alk. paper)
  1. Building.   I. Title.

TH146.B82 2005
690′.837—dc22   2005004767

A catalogue record for this title is available from the British Library

Set in 10/12 pt Palatino
by Graphicraft Limited, Hong Kong
Printed and bound in Great Britain
by Bath Press, Bath

The publisher's policy is to use permanent paper from mills that operate a sustainable forestry policy, and
which has been manufactured from pulp processed using acid-free and elementary chlorine-free practices.
Furthermore, the publisher ensures that the text paper and cover board used have met acceptable
environmental accreditation standards.

For further information on Blackwell Publishing, visit our website:
www.thatconstructionsite.com

Contents

Preface

This book has been prepared for those at the beginning of their career, studying with a view to practice in one of the professions concerned with the construction of buildings. It is written for the novice who, with study and experience, will become an expert in his or her chosen field. The more fundamental ideas developed in the text may need to be revisited as experience grows and it is, therefore, hopefully a text that can be referred to in the early years of practice and maybe beyond.

The book takes as its theme the process of choice: what the expert has to know and how the expert might think through the myriad of decisions that have to be made about the design, production, maintenance and disposal of buildings. While this involves a process of analysis, the basis of which is universally applicable, the final choice is dictated by context.

The context is set by time and place, the nature of the activity to be undertaken in the building and the aspirations of its benefactor or client. The process of choice is seeking appropriate solutions in a physical and cultural context.

For a text of this nature it may seem to be dominated by words and not illustrations. The contention is that for the novice or the practising professional faced with new ideas and situations, it is explanation of possible solutions that is required. Presenting a final working drawing may be the outcome of choice, but it is the process of analysis with its need for explanations that gives the confidence that it will perform and can be built in the context of the particular building. A picture only paints a thousand words if you already have the words to explain it.

It is hoped that the reader will find that the text tells a good story, that the development of the ideas is logical and easy to follow, but perhaps above all that it is seen as valuable in the world of practice. It is hoped that the text will prove to be illuminating and the illustrations supportive in providing the explanations of both the analysis and the solutions suggested for choice.

The first part of the book focuses on the key areas of analysis that have to be considered to check that the proposed construction will perform (will not fail) and can be built. These two tests are seen as fundamental to the final choice. In order to carry out these analyses this book suggests that it is necessary to visualise not only the physical form of final construction, however outline or tentative, but also, equally importantly, its response to the dynamic conditions under which it has to perform. The key question is 'What if I do it like this?' The answer comes from an understanding of how it works and how it is built, this understanding coming from explanations of why things happen, drawing in knowledge from a number of other disciplines such as science and economics.

Each of the areas of analysis presented in this book is independent, and therefore each analysis has to be separate, yet all have to be satisfactory before the final choice is made. Choice is a juggling act; checking a proposal in one area of analysis may suggest changes that will affect others. Structurally sound solutions may be difficult to build or solutions that will last without maintenance too expensive as a capital investment. The final choice must satisfy all the areas of analysis.

The second part of the book puts analysis into practice, focusing on making the final choice. Given the approach identified in the first part of the book, with context being so important, the chapters in Part 2 are written from the perspective of a case study. The context is the housing need for the UK at the beginning of the twenty-first century. The chapters set out to identify some alternative forms of house construction and run through the analysis to lead to some commonly adopted details and specifications. The context is set in Chapter 16 and the construction of each element of a house explored and explained in the chapters that follow.

The final chapter provides a guide to the wide range of published material that is available to develop knowledge beyond this book and particularly to support the process of choice. The book itself gives no direct references. The notion of the framework for understanding developed in the book encourages the idea of seeking relevant and current information in the context of each project. It is hoped that this final chapter introduces the sources of information that will complement the approach developed in this book in order to put these ideas into practice.

This book is written with the conviction that by focusing on the process of choice, the range of theory and knowledge that is useful to practice becomes explicit, making the link between knowledge and practice and between understanding and experience clear.

As study, and then practice, unfolds for each individual it is the power to reflect that develops understanding. It is hoped that the framework suggested in this book will help not just the process of choice but also the process of reflection. Readers should take from this text what understanding they can to use it to develop their own good judgement in whatever part that they play in the great enterprise of construction and the development of the built environment.

Part 1
Analysis

1 The Framework for Understanding

This opening chapter outlines a framework to help develop an understanding of what has to be known in order to take decisions on how we should build. The framework suggests a way of going about selecting construction and identifies the knowledge that is required to make the choice. It is the framework that will be developed and used throughout the book.

Process and knowledge

This book provides examples of construction, showing how we currently build, and provides an introduction to the understanding necessary to explain how the construction works. These two types of knowledge are both vital to making decisions as to how we should create and maintain buildings in the future. This book sets this knowledge in the context of the process of making choices for the construction. In practice, a great deal of knowledge exists of the types of construction we might use and the materials and details that might be specified. There is, however, the need for any proposed construction works to answer the questions 'Can we build it?' and 'Will it fail?' Clearly it requires the answer to these two questions to be 'Yes, we can build it' and 'No, it will not fail'. These categorical yes and no answers may be difficult to give in practice. There often needs to be some analysis to develop a level of confidence with which to make the choice. The amount of analysis required determines the level of understanding and experience needed to provide the evaluation of the suggested solution before the decision is taken to put the final proposal into practice.

It is this basic ability to make a suggestion of how a building might be constructed and then to carry out an evaluation asking the questions whether it can be built and whether it will fail that is developed in this book. The evaluation points to changes in the suggestion, which after re-evaluation will lead, through a series of refinements, to the specification and details to be adopted.

This requires knowledge of what potential solutions might look like, with current practice and precedent as the major sources for an initial suggestion. It requires an understanding of what is necessary to be specified in order to describe the proposed construction in sufficient detail to carry out the evaluation. The ability to carry out the evaluation, on the other hand, is dependent on an ability to 'see' the proposal working in the dynamic systems of the physical and social conditions in which the building is to be built.

This process of choice and the way it leads to the identification of the knowledge required is shown in Figure 1.1.

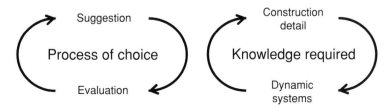

Figure 1.1   Relationship between process of choice and knowledge required.

It will be shown later that the process of choice is not just an analysis of the behaviour of physical performance. This evaluation of a proposal's response to the dynamic physical, chemical and biological systems of nature is vital, but building also takes place in a social and cultural context. There will be an imperative to ensure that the resources and know-how are available to manufacture and assemble the construction. This will require knowledge of the available industrial systems. Further, it will be necessary to ensure that the cost of the solution is monitored and that its social and environmental impact is audited. These will demand an understanding of both the economic and social systems in which the building is to be created.

The initial suggestion

If the process is suggestion and evaluation, and the starting point is suggestion, then it is necessary to know how we make the initial suggestion. How do we make that first best guess?

In most cases the suggestion is informed by precedents. It is necessary to have knowledge of current solutions and how they perform in practice. Something, somewhere, has been done before that gives clues as to how the new solutions might be formulated. In times where change is limited the particular circumstances will have been faced a number of times in the past and successful solutions will have evolved. The well-tried and tested solution needs only to be suggested and, with some evaluation to ensure the circumstances have not materially changed, can be immediately adopted. If the circumstances are changing, current solutions might still be the best starting point. They represent not only a sound basis in performance but also an existing base of resources and know-how to manufacture and produce the materials and details. Evaluation may modify but not fundamentally change the solution. This, over time, gives rise to a number of general forms from which specific solutions can be derived.

In some cases, perhaps where there are increasing user demands, or the structure of industrial practice is changing, then similar or related practice may have to be investigated. On the rare occasions where suggestions have to be derived from little or no existing previous work, a more fundamental understanding of the behaviour of the construction will have to be applied.

While experts approach making suggestions based on their knowledge and experience, it is, in many ways, not necessary to know much at all to make suggestions. It is the evaluation that requires the expertise. Novices or casual observers might make suggestions that might be hailed as brilliant observation, but in truth they have no way of knowing whether that particular suggestion is workable or not. It will take the expert to spot its potential and prove its worth through evaluation.

The power of the expert to spot potential is probably associated with the ability to carry out a rapid, approximate evaluation before subjecting the suggestion to more rigorous and explicit analysis.

Carrying out the evaluation

The heart of the success of the process of technological choice lies in the ability to be able to carry out an evaluation. The need to carry out a series of analytical exercises determines the knowledge required and manner of its application. It requires a level of understanding to answer the question 'what if' we built the building as proposed.

While the suggestion and the ultimate solution describe the construction in what appears to be a static detail, the evaluation of the suggested construction has to describe a dynamic system of behaviour.

The dynamic behaviour of the construction responding to changing conditions has to be understood, anticipating the modes of failure. Some specification of performance has to be established and then the suggestion has to be tested in the mind to assess the risk of failure within agreed design conditions.

It is easy to see technology as only the final construction, an assembly of components and materials, the building as a static object. However, making the choice of what construction should be adopted has to be rooted in an understanding of the construction as a dynamic system responding to changes in conditions and open to a failure to perform.

It is necessary to be able to visualise not only the physical construction but also its behaviour under the conditions it will have to endure in its working life. Both of these are of equal importance. If the suggestion is not visualised correctly then its behaviour under analysis may be misinterpreted. If the dynamic system of behaviour is not visualised correctly, a flawed proposal may be adopted.

Technological choice demands skills in these areas of visualising the object and the systems, and then the conceptual manipulation of the systems acting on the construction to predict behaviour and assess the risk of failure.

The criteria for choice

If the two basic questions to ask of a proposed solution are whether it can be built and whether it will fail, then the criteria for choice come from an understanding of the resources available and from the design concept as an interpretation of the client's brief. The process starts with the client's brief, the design then being derived from the requirements of those who commission the works, and as a response to the social and physical context in which the building is to be built. It is against these criteria that failure will be judged. These aspirations can only be realised if the resources are available to construct the building. This relies on the manufacturing and assembly possibilities but will also include the existence of design expertise and the options for maintenance and disposal. Knowledge of available techniques and know-how for production is crucial to the choice of the final construction if it is to be successful in reality as well as on paper.

As a design concept emerges, it is necessary to question what construction solutions might be used to fulfil the design requirements. It is then necessary to question whether this solution is available with current technology and resources within any environmental, cost or time requirements. The resources available make the design a reality. Technology stands between the design of the building and the management of the resources.

Choices can be made that may extend current production and design knowledge, but this must be recognised, and any costs involved in prototypes or training and the risks involved must be accepted before the final choice is made.

The design and resources available condition the choice. These two elements are shown in Figure 1.2 as the two areas that have to be understood before any analysis can be started leading to the final choice of construction.

Physical and social context

Figure 1.2 also indicates that before any choice for a specific building can be made it is necessary to understand something of the context in which it will be constructed. There needs to be some knowledge of the conditions that exist when and where the building is to be built, possibly with some assessment of how conditions might change in the future.

Construction takes place within, and has an impact on, the world in which it is undertaken. Some description of this world is necessary for both technical and socio-economic analyses. The world can be represented as a series of contexts, both physical and cultural. The physical context includes nature and climate. These exert forces that act on the building; they provide the raw resources and may be adversely affected by the construction and operation of buildings. The physical context includes the surrounding development and the need for the design to respond to existing buildings and spaces. People and their social, economic and political systems create the cultural context. It needs a response to local, national and even global needs based on beliefs and fundamental world views of the relationship between individuals and between society and the other components of the natural world.

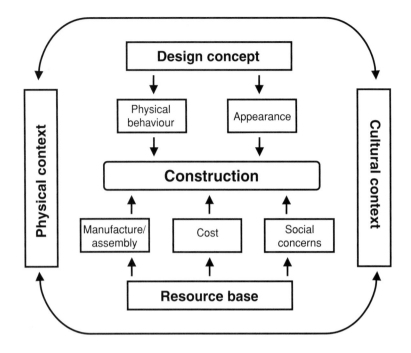

Figure 1.2   Framework of analysis.

The interaction between these two sets of contexts has been brought into focus at the beginning of the twenty-first century by the movement for 'sustainable development'. The realisation that development cannot continue without consideration of its impact on the natural environment as well as established economic and social considerations calls for new knowledge associated with materials choice, energy use and waste disposal aspects of our chosen construction.

There are, therefore, dangers in seeing these two sets of contexts as separate. However, it is still useful to consider them as having different dynamics, as, generally, in order to resolve technical questions, it is necessary to have an understanding of the physical context, while to evaluate the chance of a solution being successfully applied requires an understanding of society, its economic and political systems.

The basis of analysis

Having identified the design concept and the resource base as providing the criteria for choice and recognised the need to understand the physical and social context in which we build, it is possible to identify five areas of analysis as shown in Figure 1.2.

The design concept is the translation of the physical and social needs for the operation of the whole building into a scheme that identifies the function and performance of each of the parts. In order to achieve this, the design concept has to articulate both the arrangement and appearance of the spaces and the technical contribution each part of the construction will make to the creation and maintenance of the internal conditions.

Two tests indicated in Figure 1.2 have to be applied to a suggestion to see if it complies with the design concept:

- Does the *physical behaviour* of the construction provide a building fulfilling its functions to the required performance level?
- Does it provide the right attributes of *appearance*?

The test for physical behaviour involves three separate areas of analysis:

- Creating environments
- Under load
- Over time

Three tests indicated in Figure 1.2 have to be applied to a suggestion to see if it is achievable with available resources:

- Can it be produced, including *its manufacture and assembly* and the subsequent processes of maintenance and disposal, with existing skills and know-how in a reasonable time and at the required quality?
- Are the resources available at reasonable *cost*?
- Will it be compatible with the *social concerns* that currently exist?

These resources are both natural (from the environment) and social, so both aspects have to be part of the evaluation. It is only possible to use resources that nature has provided. However, the level of economic development in a society will provide the capacity to process materials, develop the skills and machinery to work them, the intellectual capacity to undertake design and the provision of capital to invest in the enterprise of the construction itself. Both the technical properties of materials and the existence of the knowledge to exploit them plus any environmental impact of their use must be analysed in evaluating a suggested form of construction.

These seven areas of analysis (physical behaviour being three) that have to be undertaken are, by and large, unrelated to each other and each needs its own tools of analysis and knowledge base to be successfully applied. Carrying out one analysis may indicate the need for a change in the suggestion, but the change that is chosen may make the analysis of a previous aspect invalid. It is not until all the relevant areas of analysis are shown to be satisfactory that the suggestions can be adopted as the solution.

These seven areas of analysis have potentially to be undertaken for all aspects of the construction; from the overall structural system to the finest detail such as the screw that fixes the final fitting to a wall. The work in this is clearly over-

whelming if it has to be carried out for every building that is constructed. Current practice can inform much of the process of choice. One of the most important decisions that experts take is, of all the thousands of choices that have to be made to fully specify the construction, which of the areas of analysis for which parts of the building are significant. Where, if analysis is not thought through, is the greatest risk of failure?

It is not clear how experts make this judgement. One possible explanation is that while looking at a suggestion they make many rapid checks against the seven criteria. From their knowledge and experience they take a view on what is well within failure limits and where the risks of failure are higher.

Knowledge needed for choice

It is now possible to start to identify the knowledge that is required and the areas of understanding that have to be developed in order to carry out a full evaluation. Although used in various combinations in the different areas of analysis, it is possible to put forward a tentative list of types of knowledge that will be required:

- Setting the performance levels as the criteria against which the evaluation will be made.
- Defining the conditions under which performance has to be achieved.
- Determining the fundamental behaviour of the construction that would lead to failure to meet performance requirements.
- Identifying the materials' properties that will govern the behaviour that could lead to failure.
- Thinking through the behaviour of the particular combination of materials and details under the conditions envisaged.
- Identifying the process of manufacture and assembly together with the resources required to produce the building within quality, time, cost and safety limits.
- The cost in economic, social and environmental terms of using a particular form of construction.

The process of evaluation requires the building to be conceptualised in a number of ways. While the final building will be seen as a physical construction, 'bricks and mortar', it is necessary to perceive the building in a number of other ways when carrying out the evaluation. Initially, the building has to be seen as fulfilling a set of functions with associated performance levels. For evaluation, the building has to be viewed as a series of physical systems responding dynamically to changing conditions. The building then has to be seen as a series of production operations with the resources required for the realisation of the design. All these will have economic, environmental and social implications that have to be understood. When all these can be mastered, choices can be made associated with an assessment of risk and with some confidence of a successful building.

Summary

(1) The process of choice is one of suggestions and evaluation that requires knowledge of the physical construction and the dynamic systems in which it will be built and will operate.

(2) How a suggestion is generated, and the extent to which a formal evaluation has to be undertaken, depends on the scale and nature of any changes from current practice required by either the design or the resources available.

(3) When there is a changing demand for buildings and rapidly developing technical knowledge, the role of experience changes from predominantly reproducing known solutions to the integration of experience into a process of analysis to explore the possibilities of failure in the modified or new solutions before they are built.

(4) The design and the resource availability define the criteria for the evaluation. This has to be set within a physical and cultural context.

(5) The suggestion will probably be based on precedent and this leads to general forms from which specific solutions can be derived.

(6) The evaluation needs to be carried out through seven areas of analysis: appearance, behaviour creating environments, behaviour under load, behaviour over time, manufacture and assembly, cost and social concerns.

(7) This will involve constant reference to both production and design. The concern is for what can be built and will not fail.

2 Building Purpose and Performance

Technology is used to mediate between the world as it is and the world as we want it. This chapter expands this idea, focusing on the users and their activities, introducing the functions expected of the building and how these have to be translated into a process of choice for the construction of each part.

Activities, space and construction

Buildings are often characterised by their overall purpose or function – house, factory, hospital, prison, etc. – and are often immediately recognisable although their architecture may vary. To understand the form of the building, it is necessary to understand the day-to-day activities that make up the function for which the building is commissioned. The construction will have to play its part in ensuring the well-being of these activities. The notion of well-being is normally associated with the people who use the spaces, but the building will have to protect the equipment being used and the storage of items associated with the activities. Well-being of people also needs to include the requirements of the owner and the wider public as buildings are social institutions and create the surrounding spaces.

The majority of spaces in buildings have to accommodate people and consideration of their well-being becomes a dominant aspect of the performance of the construction. In some cases, such as the rooms for the early computers and specialist storage areas such as vaults and cold stores, the equipment housed in the building or the storage conditions dominate the requirements. Designing for the well-being of people is essentially concerned with health, safety and comfort. Health should be seen as both physical and mental health which is recognised as being a function of both physical and social conditions created by the building.

The owner may see a purpose beyond the simple support of activities, to include image and status. The surroundings in which the activities are undertaken have social significance and must also form a feature of the design.

Each society over time evolves common building forms for each purpose type that become recognisable within that society because of their size, arrangements of internal spaces and social statement. With this emergence of building form there is a need to develop appropriate construction solutions. This process is dealt with in more detail in Chapter 3, but first this chapter explores the complex range of performance issues. These will be established predominantly by the users' activities although later by the owner and then by government in the form of legislation on behalf of society as a whole.

Introduction to performance requirements

In order to explore the range of performance issues it is useful to engage in a mind experiment. The situation to be imagined is one where civilisation as we know it does not exist and the task is to provide shelter. The scenario is one of few resources of materials, labour or knowledge. There is also little development of an economic system so the whole design and production process has to be undertaken by the users. The mind experiment starts by asking which performance the user of the building would seek to satisfy first. The experiment can then be extended through time, assuming increasing resources and socialisation, introducing additional requirements that would add value and utility and improve well-being.

By undertaking this mind experiment it will be possible to introduce the needs of the user and hence the functions that he or she will require the building to fulfil. As the user is also, at least initially, the designer and builder of the shelter, he/she must decide on the construction. The experiment will, therefore, also start to introduce the ideas and knowledge that will be needed to make decisions on the construction of the building itself. All the issues and factors introduced here are still relevant today.

In the beginning

Given no immediate risk from attack we can speculate that people would organise their technological endeavour in the following way. Thought would have to be given to the range of activities of everyday life that should be housed in the shelter. Initially this is likely to be very few, perhaps only sleeping, with most activity being undertaken outside depending on the prevailing climate. This will provide a definition of size and any division of the internal space that needs to be made. It would be immediately recognised that these requirements for size and shape will be limited by the possible construction available. Some decision would have to be made as to how much effort should go into developing construction to match aspirations or use existing resources to satisfy immediate needs.

The priority in terms of establishing functions for the building fabric would be to remain dry, a condition fundamental to the health and comfort of people and the storage of many natural products. This would be accompanied by the requirement to be warm, but it is possible that this may initially be achieved in combination with other technologies. These would include clothing and the use of fire which, at this stage, may or may not be associated with the construction of the shelter.

As soon as any construction is contemplated the builder is immediately faced with achieving a stable structure and making some judgment on how long it might last. Although buildings may be erected for environmental control, their very existence demands considerations of stability and reliability if they are to be deemed satisfactory by the user. This will require the identification of possible modes of failure. At this early stage this would include how the construction might leak or fall down. This understanding of the ways in which performance may be lost becomes crucial to directing effort to significant aspects of the solution. This would also lead to some questioning of what maintenance may be necessary and what parts may have to be renewed to continue the performance until the construction had to be abandoned or demolished to make way for a new building.

It would also become necessary to have some knowledge of the environmental conditions in which the construction has to maintain performance. This knowledge would be required to consider not only the environmental control but also the stability and reliability. These three become the key physical performance issues. External climatic conditions such as frequency and volume of rainfall will need to be assessed for the weatherproofing function. Damaging wind speeds and direction would need to be assessed for stability. It would, however, soon become clear that wind is a major factor in weatherproofing. Some of the most arduous weatherproofing conditions come with wind-driven rain. It is not possible to say what aspect of the environment needs to be known to assess durability until some choice of the basic materials is made. Specific decay agents in the environment degrade different materials. Linking materials to their generators of decay is just one more question that has to be answered in the quest for knowledge to make sensible choices for the construction.

This assessment of external climate would soon reveal that often the more extreme conditions of open land are ameliorated by local natural features; that exposure can be modified by the position and orientation of the building which will limit the demands on the construction and/or give better internal conditions.

The next steps

As the number of activities carried out in the building increases, the provision of good air may be the next priority for improvement of the construction. This is probably the next requirement for health and comfort after dryness and warmth.

Seeking a solution that would improve the quality of the air would probably involve ventilation since there is an abundance of good air outside. It would have to be recognised that the air outside may also be cold and that will make the requirement for warmth more difficult to achieve. This may mean that a solution for good air may only be possible when air temperatures outside are sufficiently high or air quality inside is particularly poor. Assuming that the active technologies of building services are not available, modification would be necessary in the existing construction. It would be necessary to use natural ventilation driven by wind pressures around the building. Openings would have to be introduced to ensure the exchange of air. Their position and size would have to be considered. It would be possible to introduce a level of control by providing shutters. The openings could eventually be as sophisticated as chimneys, but history tells us that these often come much later.

It is important to note that the containment of the bad air is a direct result of employing construction to remain dry and warm. The air outside the building is almost certainly good. This realisation that the construction itself can create undesirable conditions should not go unnoticed. Changes introduced to fulfil new performance demands may change conditions that have always been satisfactory. This happened towards the end of the twentieth century when a requirement for energy saving was satisfied by the introduction of insulation. Under some environmental conditions this led to condensation that has a deleterious effect on the basic requirement of keeping dry. This led to health problems for the users of the buildings.

The introduction of new performance requirements may also introduce failures in the construction that have previously not occurred. This happened in the condensation example where dampness caused not only health problems but also a deterioration of the construction materials themselves. In this mind experiment it is possible to speculate on a similar problem with the improvement in internal air quality. Given the lack of social organisation to produce more processed materials the construction would almost certainly be made of relatively unmodified organic materials. In this case it might be speculated that if the poor air had a large component of smoke it would have permeated the construction. This would have effectively fumigated the organic materials and kept infestations or decay-promoting fungi and insects at bay. Improving internal air quality may change the conditions within the construction itself. This may allow significant numbers of some of the natural agents of decay to thrive, thereby reducing the life of the construction.

More physical requirements

This mind experiment continues on the assumption that no major threat to security arises. This would inevitably lead to a diversion of resources and may change the priorities of individuals or society in seeking improvements in their buildings, although it may also lead to new types of buildings such as defensive works.

Given peaceful social development, the next priority may be difficult to identify. It may be that the infestation of insects and small animals that would be attracted to the improved internal environment may be becoming a problem. It may be that in the harsh realities of life, and in the absence of any realisation of the health risks, these infestations would be tolerated. Measures to eliminate the larger animals, if not already considered as a danger, may now become important. At some time, however, infestation control would have to be included in the list of functions required of the construction to improve well-being. This is part of the more general requirement to keep the enclosed space clean. Internal surfaces that are dust-free, smooth and wipeable become desirable attributes where cleanliness becomes important to health and comfort for the user.

Having gained good clear air, more activities could be carried on indoors. It is, therefore, possible to imagine an increasing need for light to improve comfort levels and further extend the range of activities in the building. The choice is between natural and artificial means of lighting. Natural would involve openings, while artificial would only demand a safe place to support what would almost certainly be an independent light source. Although independent of the building, it would have an influence on it. Perhaps the major influence it would have is in the increased risk from fire. While it may not be possible to improve the fire resistance of the whole building at this time it would be sensible to employ non-combustible materials at least where the light source is to be mounted.

Here we see the increased use of technologies independent of the building, seen by the user as an asset to the well-being of the activities, influencing the required functions of the construction itself.

Another requirement is for some measure of noise control. This would, of course, only become a concern if the external level of noise were somehow interfering with the overall sense of well-being. It may become a problem when the building has separate rooms for specific activities. Noise from one activity may well be incompatible with the conditions necessary for another. It would seem that noise is a particular problem of civilisation, particularly those who have advanced technologies. It also seems that it has a cultural basis in attitudes towards individual rights, privacy and intrusion.

The technical considerations become more complex

It is useful to view the need to consider noise performance from a different perspective. It may be that there is a need to modify the sound between outside and inside, yet the construction used for the other requirements already has sufficient, if not excess, ability to perform as a sound barrier. With low performance expectations the final form of construction is dictated by only a few of the functional requirements. As more functions are identified and the level of the required performance increases it becomes difficult to provide simple construction solutions. Composite solutions are necessary with different materials to satisfy different functions at the levels of performance that are required.

The analysis of exactly how each part of the construction is interacting in providing the overall function becomes more difficult. The danger comes when changes are suggested. It is necessary to ensure that the change does not make a previously satisfactory performance critical so that the construction fails to perform, in an unexpected way. The trend towards lighter partitioning in housing has, by reducing the weight and changing natural vibration frequencies, reduced the sound transmission resistance. In the past this was normally more than satisfactory when partitions were thicker or made of denser material. The lighter, cheaper solution may become critical in some situations where the soundproofing performance of the heavier solution had always been taken for granted.

The need for composite construction leads to the search for materials with properties matched to the specific functions. Processing or manufacture of materials allows these properties to be more accurately achieved. This not only changes the ability to gain performance but also influences the production resources required. Materials and manufacturing methods may also be changed to provide more economic buildings. The innovation may not be driven by performance but in order to seek production and cost gains. This may lead to performance failure in previously satisfactory solutions. Those responsible for the choice of construction must re-evaluate the possible performance failures if construction is changed for production or cost advantages.

Taking storage and equipment into account

The environmental requirements that have been introduced are those necessary to maintain the well-being of the people who use the building. These can be summarised as:

- Dryness
- Warmth
- Cleanliness
- Light
- Quiet

Observations of this list will identify that, with only slight modifications (e.g. warmth needs to be more generally interpreted as temperature control), it could equally well be applied to maintaining the well-being of the equipment and stored items. The required internal conditions for storage and equipment may be different, even hostile to life as in the case of cold storage. They will still demand the same type of functions of the construction. There would, however, be changes in performance level due to the change in specification for the internal environment even though the external condition may be the same. In the example of cold storage, levels of insulation would be different.

More significantly, the equipment and stored items, as well as the living occupants, will contribute to the internal environment with noise, smell, heat, fumes, etc. It may be that the construction is called on to moderate these effects by isolating storage from living areas by partitions. Independent technologies in the same room such as cooling of food may be applied using ideas such as evaporating water from a porous pot or a refrigerator. However, independent technologies may influence internal conditions. In the example of cooling food the porous pot releases vapour and the refrigerator releases heat.

The dynamics of these changes in the internal environment due to occupants and independent technologies in the building must become part of the analysis. Sealing up a building to reduce heat loss through air changes can lead to a buildup of vapour (if only from people breathing) which makes condensation more likely.

Although the list of environmental requirements generated for people will cover most circumstances associated with the equipment and storage in the building, it may in some situations be incomplete. Some modern equipment

has environmental operating requirements such as the need for anti-static precautions that may not currently be considered necessary for the well-being of people. What has to be achieved is a balanced solution to ensure the well-being of both the participants and the processes associated with the activities for which the building is required.

What should be considered tolerable?

Setting performance levels involves considering what is tolerable. What should be considered as normal limits to be maintained in the internal conditions? What external conditions should be allowed for in the design? These questions have to be answered in terms of risk and consequences. Extremes that happen rarely but have serious consequences to the well-being of the activity or even threaten life and property may well be taken into account. The more frequent events, even with lesser consequences, are also more likely to be taken into account. This is not independent of the economic resources available and social attitudes. In many parts of the world, earthquakes will occur, but the construction to minimise the risk to life is seen as increasing costs beyond that to be extended to all sections of society. Similar circumstances have existed in the past with fire and building. Increasing wealth, effectively distributed in society, will affect what is considered tolerable, reflecting more demanding performance expectations.

Generally, wide limits would be accepted for solutions at the beginning of our mind experiment, but as societies develop they seek to extend the range and level of demands. This often includes a requirement for a greater level of control as it defines a narrower band within which performance is deemed satisfactory. These trends place more demands on the solutions and failure is more likely.

Some extremes that happen rarely have to be considered if they have serious consequences. One risk that comes from the internal activity that generates extreme conditions has already been raised. It is the issue of fire within the building. At the beginning of the mind experiment, while escape for people would be required, it is unlikely that there would be any expectation on the part of the user that any of the building would remain after a fire. Although fire is now rare in modern buildings, its consequences are such, particularly in risk to life, that it is currently deemed necessary for the construction to continue to maintain a selected range of performances both during and after the fire. This would almost certainly include stability, although insurance requirements may extend the performance range.

External climatic extremes such as wind, flood, lightning and even earthquake may be considered. Design for flood may be sensible if the building is sited by water, for lightning if a tall building and for earthquake if the building is in an unstable region of the world.

New threats created by increasing sophistication of solutions are now emerging. Draught-free homes created to conserve heat may suffer from a buildup of radon gas if built in an area of the country where it occurs naturally in the rocks below the building. The chemical contaminants in the soil on sites used for redundant industries threaten both the occupants and the fabric of the building itself.

Introducing social requirements

Having established the requirements for health and physical comfort it is now necessary to consider the social aspects. In reality the consideration of performance for more social aspects does not come only when physical requirements have been fully satisfied. Human endeavour very quickly becomes a social activity and there grows a need for social organisation, establishing cultural norms, provision for leadership and political activities and an economic system for the exchange of goods and services. Buildings play their part in this process not only to house the activities but also in that they reflect, affirm and even become instruments for change as the cultural and economic order develops.

As well as the issues of status and image, to be discussed later in this chapter, two other functions associated with well-being arise when society starts to develop. These are the requirements for security and privacy. They are mainly associated with mental health, although the security risk, if it becomes a reality, may involve physical danger. The point at which people would take these functions into account would be dependent on the nature of the society that develops. They are not as predictable as the physical needs and may well have to be addressed before the desired level of physical comfort and safety is achieved.

Security requirements will be dependent on the existence of a perceived threat. If the risks are seen as sufficient then technological ingenuity, time and resources will be devoted to combating the threat if social devices such as the rule of law or treaties cannot be achieved. Threats come from both inside and outside a society, each of which pose different challenges to the technological skills of the builder. In the extreme, in buildings such as bunkers, the security threat dominates the design as the technology of survival becomes pitted against the technology of destruction.

The requirement of privacy is much more rooted in the cultural attitudes to community. It is cultural interpretations of what activities should be shared and those that should be carried out in private that dictate the need to introduce this performance into buildings. Performance levels of privacy can be identified associated with the senses, visual and aural being the two main requirements of the building influencing both the external envelope and normally demanding some division of internal space.

The functions of the building that have a social origin often translate into aspects of physical performance already introduced. Security makes structural demands on the fabric. Privacy determines many of the sound transmission requirements of construction.

Influence of production resources

As more functions are included and more materials are introduced, production resources and know-how have to develop. The relationship between the need for a building and the availability of the materials, skill and tools to produce it is always a determining factor in the choice of the construction. Access to production knowledge and expertise limits the pace of development of new solutions to achieve new performance. However, developments in materials and production can also suggest different ways of achieving similar performance.

This mind experiment has taken the perspective of the purpose of the building and requirements of the user. It is easy to forget that the means of realising that performance may need as much thought and ingenuity as the design and detailing of the fabric and structure of the building. In many old buildings it is easy to see how they work but often difficult to see how they might have been built without appreciating the tools and skills available at the time. In many innovative buildings, detailing is often determined as much by manufacturing or erection procedures as it is by satisfying performance. The availability of manufacturing and assembly resources also determines the cost. The need to develop production procedure affects the time to construct and risk associated with a particular solution. All these contribute to the final quality of the building and its success judged by performance.

Considering cost and value

Throughout this mind experiment the question of cost and value will have to be considered. Buildings are usually costed in terms of money, but there are also considerations of cost to the environment and on the impact they might have on the society itself. The resources available to any society are limited and decisions have to be made as to how these should be allocated. As part of the culture these are the political decisions involving both the vision of the type of society that should emerge and the preferred methods of achieving it. It is from this cultural process that demand is initiated, the cost of resources determined and value assigned.

Buildings will only be commissioned when the cost of a building that performs to an adequate standard is considered to be good value. If the costs are too great then decisions have to be taken on eliminating functions, reducing performance or changing production methods. These changes have to ensure that the building is still considered as value for money. Assigning this cost requires that at least a broad outline of both the construction and the production methods is known. At some point in the process the decision to proceed to full design and production based on estimated cost has to be made. For buildings that can be produced using well-developed current forms of construction this decision can be made very early with some certainty. For a project where there are few precedents, possible solutions may have to be investigated before cost can be established with sufficient certainty to take the decision to proceed. It is, however, always a requirement to set cost limits as early as possible and then develop the final solution with these costs in mind.

Building for status and image

Status and image will at some time in the development of society start to influence the requirements of the building. This may be reflected in the size of the space provided over and above that strictly required for the physical satisfaction of the activity itself. It may be associated with the quality of the materials used being acknowledged as expensive. These expensive materials must have the ability to perform their physical tasks without failure, but they could have been replaced by a less expensive alternative. These requirements are often then overlaid with issues of art and fashion and symbolism. The influence of these should not be underestimated, but a full discussion of their nature and form is inappropriate in this text.

The pressures of status and image are not just related to statements of wealth and position. We all need to make statements and feel comfortable about our place in society related to the prevailing cultural norms. It has to be recognised that these cultural norms can be highly regionalised, being associated with tribes or nations, although improved communication seems to reduce this variability. We all care about what our buildings look like and the environment they create. Using and being associated with buildings that match the psychological demands of the activities carried out in the building is very important to our mental well-being.

These issues become associated with aesthetics, the visual and deeper, even spiritual, quality of beauty of both natural and manufactured articles. These are the aspects of the world that satisfy our feelings rather than our physical needs. Buildings intrude into the rural landscape and create the cityscape. Their shape, form, colour and ornamentation have a significant impact on our lives.

Control from society and legislation

All social groups have cultural norms of behaviour that limit the actions of individuals and impose sanctions for transgressions. This eventually has to be formalised into legislation, some of which will start to impinge on the choices that may be made for construction. Direct legislation is normally associated with public health, limiting the use of scarce resources or the position and appearance of whole buildings. Once manufacture and assembly start, there are laws associated with the health and safety of operatives and people in the surrounding area.

All these will have to be acknowledged and satisfied in the final choice. If the choice of construction is made in the sprit of responsibility within the cultural norms it should not dominate the solutions chosen. Legislation is for the protection of those who use the buildings and should, therefore, be easy to accommodate into a process whose objectives include satisfying social and individual needs.

In reality the law on these matters can be slow to change and may be drafted such that its interpretation becomes a matter of opinion. These factors can spur or inhibit technological change, but it is clear that in an advanced society those responsible for the choice of construction have to be aware of the law and its interpretation.

There is much legislation that will indirectly influence the choice of construction, although predicting and understanding the links are more difficult. Contracts become necessary between the organisations that undertake different parts of the building process. This legislation affects decisions on the construction and what risks they carry. The knowledge and experience that the organisation has then influence the choices made for the construction of the building.

Clients, users and change of purpose

Another complication that arises as society develops is that individuals who become the owner-occupiers commission fewer buildings. It is now possible for those who commission the work to be an investment company who pass on the task of operating the building to another company for use by yet another group of people. This does not nullify the idea that satisfying the needs of the user should govern the performance of the construction solutions. It is the provision of space that ensures that the desired activities can be carried out efficiently and effectively by the user that gives value and utility to a building. Those who commission the building may have objectives such as investment, but these will generally only succeed if the purpose of the building and its performance are satisfactory.

Even when relatively simple relationships between client and initial occupation exist, the purpose of the building is likely to change with time. These can be quite dramatic changes such as churches to flats, but most are much more subtle. Changing practices in medicine have made many old hospital designs inefficient and the general electronic information revolution has changed the basic activities carried out in many offices. Increasingly the operators of buildings (as opposed to the users) see the building as part of the business efficiency. This gives different value to maintenance, alterations and refurbishment and makes the building a more integrated part of the facilities as a whole. This leads to the questions about the whole life costs. More expensive initial options may be better value over the life of the building, reducing operating costs and increasing efficiency and well-being.

Environmental impact and sustainability

This long-term view of the building leads to wider questions about environmental impact. Construction has always had a local impact on the environment; it takes resources, changes ecosystems, consumes energy and creates waste. The society in both our mind experiment and now in reality has grown more prosperous; infrastructures have improved, allowing the growth of manufacturing and materials processing to serve an expanding population with ever higher performance expectations. The environmental impact of technologies chosen for individual buildings can now contribute to global consequences.

This demands yet more analysis, with new knowledge required in order to make choices about the technology we employ for our buildings. It will lead to different solutions being favoured, innovation and risk. The story continues.

Summary

(1) The purpose of the building is to support the activity to be carried out in the building using both physical and social criteria.
(2) Both physical and social functions of the building lead to physical performance requirements covering environmental, structural and reliability aspects of the construction. These requirements lead to performance expectations and hence definitions of failure.
(3) Choice of construction can only be made in the knowledge of the environments in which the construction operates and an understanding of the mechanisms or actions that lead to failure.
(4) The other factors to be considered in making the choice of construction include production resources and know-how, cost, legislation and the need to consider the whole life of the building and its environmental impact.

3 Common Forms – Specific Solutions

This section develops the idea that in any society there is a recognisable set of common building forms. These common forms will be drawn from a range of generic types that are then worked into various specific solutions. The generic type and constructed form are two distinct ways of identifying the construction that is adopted for a building. The conceptual generic type is characterised by its arrangements of components and action in achieving performance. The physical constructed solution has defined materials, size and shape of components, joints and fixings. This construction solution becomes specific to an individual building, although the building may be repeated many times and the detail may work in many buildings.

General and constructed forms

In any society at any time there is a pattern of needs that supports a recognisable series of common building forms. For each of these building forms there will be an interrelated set of construction solutions to cope with the size of the building and the resources affordable for each. Construction solutions are derived from a limited set of generic types. Technological choice often starts with the selection of an appropriate general form that is then developed into a detailed construction solution for a specific building.

The ability to describe a building in these two forms, the general form and constructed form, at the early concept stage in the design of a building is most important. It should be the big idea from which the individual details can be developed so they work together to ensure the performance of the building as a whole.

The emergence of general building forms

It appears that the prime determinant of the construction of a building is related to space. The users' demands will translate fundamentally into spaces and their size and interrelationships will suggest the scale and shape of the building. At the domestic scale small cellular spaces lock together, while many commercial or public buildings require more voluminous spaces, often with smaller associated spaces such as toilets, to form the building as a working unit.

Construction solutions are directly related to this demand for space. Most significantly it influences the structure, but these choices have inevitable consequences on the solutions available for the construction of the enclosure and other elements of the building.

Currently there is a wide range of spatial requirements both on plan and in the height required that leads to the use of a range of structural forms. Three general groups of structural form can be identified:

- Load bearing walls
- Skeletal frames
- Long span roofs

Each of these will create a range of building forms, some examples of which are shown in Figure 3.1. Each building form will be made up of a combination of a range of structural elements. An example of a generic structural element is the beam. With its structural action in bending it appears in cellular loadbearing construction as joists for the floor and a lintel over an opening, in skeletal frames as a main structural member and in the long span roof in a portal frame. This generic type has been adapted for specific solutions using different materials

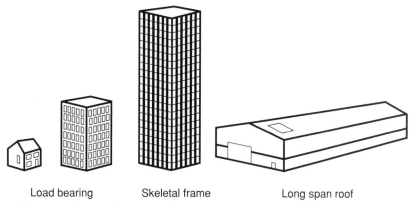

Load bearing   Skeletal frame   Long span roof

Figure 3.1 Typical form associated with types of structure.

to achieve the required characteristics: great strength, small deflections, small section size, low weight, good looks, low cost, fast erection. Each of these characteristics will have a greater or lesser significance in a specific building. In a house the joist will most likely be timber, in the skeletal frame the beam will generally be steel or reinforced concrete. In the portal frame it is most likely to be in steel, but for a building of the right use, scale and size it could be made from reinforced concrete or even from timber in a laminated form.

While the ability to enclose the space is the primary determinant of the generic form, design and resource conditions will lead to a dominant economic detailed solution. This leads to two ways of thinking about a building, each having it own vocabulary to describe the solution. Early in the selection of the construction it is helpful to think about the elements of the building in generic terms. This can then be developed into a full description of the constructed form including specification of materials, joints and fixings.

Generic elemental – domestic, loadbearing walls

One widespread use for the loadbearing wall form is the house. Described in their elemental form, houses in the UK at the beginning of the twenty-first century employ a semi-permeable (cavity) construction for all external walls to provide (with internal cell walls) a stable dry box. This box will be roofed with a tiled covering supported on a trussed rafter structure taking the loads directly to the external walls. This means that any internal loadbearing walls need not extend to roof level. Suspended floors will be joisted, often taking advantage of internal walls to limit spans, making some ground floor partitions loadbearing. Ground floors may be ground supported or suspended floors depending on the site and ground conditions. Foundations are most likely to be strip footings unless the ground conditions are very poor. This form is shown in Figure 3.2.

Many houses built in the first half of the twentieth century and still in use were not constructed like this. Many have solid permeable walls and have roofs made from individual structural members (rafters, purlins, struts and hangers) which, when acting as a whole, require internal wall support. Later in the twentieth century the width of terraced houses reduced and cross wall construction, where only walls between the houses are loadbearing, became an option. This lost the stabilising effect of the box but allowed a range of less expensive external back and front walls to be chosen.

As the loadbearing wall solution is developed into multi-storey flats other elemental forms have to be employed. Floors with greater sound and fire performance requirements become more economical as slabs, not joisted, even though the spans may not have changed significantly. The loadbearing walls may become prefabricated,

Figure 3.2   Generic description of house construction.

**Figure labels:**
- Tiled pitched roof
- Trussed rafter roof structure
- Non-loadbearing partition
- Joisted first floor
- Internal wall carrying floor loads
- Semi-permeable cavity wall
- Ground-supported floor
- Strip foundations

taking advantage of the repetitions floor by floor to gain a production advantage.

As this arrangement is developed into multi-storey blocks, the alternative structural form of the skeletal frames will become an economic alternative, even though the plan layout still shows walls dividing the building into small spaces with limited floor spans that made the loadbearing wall option viable.

Constructed form – domestic, loadbearing walls

None of the generic descriptions talks about the material, the shape or size of the components or the joints and fixings employed. They give the elements general names and describe them in terms of their general form or how they perform. For the constructed form, details have to be worked out to ensure economic sizes and manufacturing techniques. From this a recognisable physical constructed form will appear. The

constructed form of the generic elements shown in Figure 3.2 is given in Figure 3.3. This shows concrete single lap interlocking roof tiles supported on timber battens over felt fixed to timber-trussed rafters jointed with galvanised steel gang nailed plates. The external walls are masonry although timber frame is a viable alternative. The wall may be constructed of brick outer skin and block or timber framed panels for the inner leaf. This is unlikely to meet the requirements to limit the passage of heat. The most cost effective way to limit heat loss in the winter condition (loss of heat from warmed inside air) is by insulation incorporated into the wall. In summer the design condition is heat gain from outside. As this gain is predominantly from direct radiant heat from the sun, shading or thermal capacity (not shown in Figure 3.3) may be better actions than insulation.

The floors are timber joist supported by resting either directly on the inner leaf or on galvanised steel joist hangers and finished with chipboard

Concrete single lap tiles fixed to timber battens over sarking felt

Timber truss created with galvanised steel nail plates

Timber stud partition with plasterboard skin

Timber joists bearing on inner wall with chipboard decking and plasterboard for ceiling

Brick outer leaf with block or timber frame inner leaf connected with wall ties with insulation in cavity

Inner block wall with plaster or plasterboard finish

Solid ground floor with fill (hardcore) under concrete slab

Plain concrete strip foundations

Figure 3.3   Constructed form description of house construction.

sheet flooring. As the ground floor partition walls carry the joists they are constructed with concrete blocks. However, as the trussed rafters only transfer load to the external walls the first floor partitions only need to be timber stud with plasterboard finish as long as they can provide appropriate levels of sound insulation. Plasterboard has been used for the ceilings and as a finish for the downstairs blockwork, although gypsum plaster may be used on any block walls in the house. The external wall would have to be modified below ground and the foundation would be a strip of plain concrete.

The size of each of these components has to be chosen. Common components, such as the brick or the standard range of windows, have been developed in appropriate sizes to work with the domestic scale and performance levels. The size of some components, such as the trussed rafters and floor joists, has to be designed to the size and use of the building.

In the same way as houses in the past have been different to the construction shown in Figures 3.2 and 3.3 there is no reason to suspect that either generic forms or the physical solutions will endure into the future. They will evolve and develop to meet new design needs and resource conditions.

Other general forms and current practice

Buildings using skeletal frames and long span roofs show a greater variety of both generic elemental types and physical construction solutions. To try to catalogue them here is not appropriate. However, a few examples serve to introduce these forms as both generic and physical entities.

The skeletal frame structure has beams, columns, slabs and some means of ensuring overall stability. This basic format can be repeated to generate large buildings both on plan and in height. These structures will, however, need internal structural columns or walls, thus imposing some limits on clear internal space. Where this structural form has no current rivals is in its ability to achieve great heights with relatively open internal spaces. The physical constructed form is generated in either steel or reinforced concrete. The reinforced concrete can be either precast or in situ. These materials can then be mixed between the elements. Most steel solutions have steel beams and columns but concrete floors and either concrete or steel wind-stability provision. Reinforced concrete structures normally have reinforced concrete members throughout, but the development of flat slabs as spanning structures has eliminated the need for beams.

Long span roof structures are used for industrial, retail, storage, exhibition, sports venues and other activities that need large floor space under the roof. They are normally only single storey, although they may require considerable internal clear height. Some buildings used for these purposes are not large so although termed long span because of their potential, the generic forms may be used for buildings of modest scale. Long span roofs have perhaps the greatest variety of generic elemental forms to be developed into construction forms. Both plane and three-dimensional structures in flat and curved forms have been developed in steel and reinforced concrete and even timber and aluminium.

Both skeletal and long span structures have to be enclosed. The enclosing fabric can normally be supported and restrained by the structure and there is a wide range of options for the external walling. Again generic forms such as cladding or facings can be identified, with constructed solution in materials such as metal, concrete, glass or brickwork being chosen for the specific building. The different structural forms tend to use different enclosure solutions. In long span structures the building often requires less openings, and reducing the weight of the fabric allows economic gains to be made in the structure. This leads to sheet cladding in metals with heavier forms such as brickwork used only around the building at ground level where it can be supported directly on the ground. In the multi-storey skeletal form the interaction of the enclosure and the structure plus the need for performance often lead to more complex construction.

Uncertainty and risk

Where change is slow, there is the opportunity for current detailed solutions to be adopted with little modification from one similar building to another. There is always danger in assuming two building to be the same, but if the general form is robust and not being applied close to a failure cusp the solution can be suggested in detail without much analysis, with little risk.

As the pace of change increases, initial choice may deviate from currently detailed and even generic forms. The initial suggestion becomes based on an understanding of generic forms that then need to be thought through in detail. Initially this may only need some alterations in materials or detailing, but as demands become more divergent from common practice then alternative elemental generic forms in new combinations may be more appropriate.

In making any technological choice it is important to be aware of the uncertainty and risk in adopting a particular solution. The technical risk is one of failure in performance. As with any risk it must be weighed against the consequences. Consequences that lead to loss of life always require careful attention. Where the consequence is interruptions in the operation of the building, the severity of the loss will have to be evaluated against the risk of failure.

Whatever the pace of change, the choice of construction is based on an understanding of the action of the generic form and the performance of the detailed design and availability of resources to construct the specific solution.

Summary

(1) Elements and components have a limited range of conceptual generic forms based on their action in achieving performance.

(2) Detailed construction solutions have to be chosen for specific design and resource conditions and are therefore ultimately specific to an individual building, although the building may be repeated and one detail may work in more than one building.

(3) Within each society a range of common building forms and economic construction solutions arise which normally form the basis for choice for subsequent construction solutions.

(4) Particularly in times of change the process of choice has to include analysis that recognises the uncertainties and risks for each individual building in both failing to achieve design performance and acknowledging the reality of available resources.

4 Construction Variables

This chapter explores the variables that define construction: materials, shape, size and spatial relationships. It introduces the reality of deviations in assessing whether the components will fit together and perform. The chapter concludes with a section on the specific aspects of jointing and fixing as the process by which the parts of the building can contribute to the behaviour of the whole.

The variables of construction – the outcome of choice

It is important to recognise at the start what aspects of the physical construction can be varied in order to achieve the desired performance of the building. Identifying what can be varied also identifies what has to be chosen and specified if an unambiguous solution is to be defined.

Choices have to be made for the materials to be formed into components that have a defined shape and size. Each component has then to be given a spatial relationship with other components together with the joints and fixings necessary for them to act as a whole building. This simple identification of the variables is represented in Figure 4.1.

Defining the final physical arrangements of the construction involves a specification for each of the component parts of the building, including any joints and fixings, detailed in assemblies of components to create the final building. This specification will have to include:

- The material of which components are be made and the quality of that material
- The shape into which the material has to be formed
- The size of each component giving dimensions sufficient to define the shape
- The relationship of each part either to an absolute position defined by a reference framework or relative to another part of the construction or the surrounding environment

This specification determines the space that the components and their joints will occupy. It may be in competition for this space with other parts of the construction and/or the useable space of the building itself. The choice of construction cannot be confirmed until this has been checked and it has been established that there is a unique space for each part of the building.

Joints and fixings also need to be identified by their material, shape, size and spatial relationships, as they themselves are components in the assembly. These assemblies are then joined together (often identified as junctions) into the whole building. Any components associated with junctions also have to be specified.

The process of choice is about defining the materials, component shape and size, spatial relationships, and joints and fixings necessary to

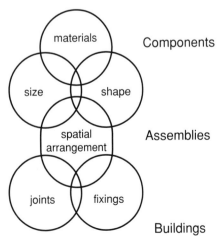

Figure 4.1 Variables or outcome of choice.

complete the building. The focus of technological knowledge is taking decisions about these variables to create a building that is safe and has utility. This is a technical choice that has to be made following the creation of a design concept and coming before the management of the production process. Yet it is a choice that has to be made in the full knowledge of the design requirements and the available production resources.

Choosing materials

Specifying materials is a process of matching the properties of the materials to the performance requirements. This is not, however, independent of size and shape, for all materials have all properties to some degree or another. We only choose to exploit certain properties when their value is such that, in economic quantities and at sizes and shapes that can be produced, transported and fitted within the space available, they will perform at the required level.

Buildings are perhaps the largest manufactured items widely used throughout all societies in the world. They demand relatively high quantities of materials at reasonable cost and it is, therefore, not surprising that, historically, materials used for everyday buildings are those with high bulk, locally available and with limited process costs. Increasing economic wealth and its attendant higher levels of infrastructure allow the use of materials requiring greater processing off site and less need to rely on local materials. With economic development comes the possibility for transfer of production activity to factories remote from the site, which in turn gives opportunities to use a wider range of materials than a wholly site-based process.

The materials that have been dominant in building in the past are timber and stone, with lime, limited amounts of iron for fixings, lead and glass. Fired clay for bricks and tiles has been used for some time in the UK, but up to just before the turn of the nineteenth century these would have been only used locally, particularly in the conurbations where there was a sufficiently large market close to the kiln site. In rural areas such as Devon, soil was still being used for agricultural cottages up to the end of the nineteenth century.

Perhaps the two materials that have made the greatest impact on the size and scale of everyday building have been steel and concrete. The superior strength properties have allowed greater loads to be carried on smaller sections but in particular have introduced the possibility of smaller, efficient spanning members such as beams. This opened the way to large, comparatively voluminous building to a wider group of building clients. In the last half of the twentieth century there was an increasing number of materials introduced into buildings. More metals such as copper and aluminium now are used, particularly in services. Plastic has made a considerable impact but only where its relatively high cost can be justified by thin sectioned components. Processing and manufacturing developments have offered some old materials new opportunities with techniques such as laminating timbers.

As well as increasing the range of materials, developments have been made that offer more grades with greater quality control of the more traditional materials, allowing even finer choices to match performance with cost. All this demands increasing knowledge. Evaluating the more recently formulated materials is more difficult for many have not yet stood the test of time. More materials introduce more possibilities for incompatibility so even well tried and tested materials may fail in new combinations. Adverse reactions between materials and between materials and the environment may threaten durability in ways that have to be imagined before an analysis can be undertaken.

In imagining failures it is possible to argue that all failures are ultimately materials failures. It is also possible to argue that the failure of the material is always the consequence of a previous sequence of events and therefore never the direct fault of the material. The usefulness of this argument possibly lies in accepting that both are true. In considering the behaviour of the material, it is necessary to identify the significant events in the life of a component when a coincidence of conditions may lead to failure.

The events that lead to failure will determine which properties of the material will become critical. Significant events may well be during manufacture, assembly or during the life of the building. Understanding the production process as well as the behaviour of the finished building is vital, for the most demanding event in the life of a component may well be in storage, handling or as part of a partially erected building. An example of this is a precast, reinforced concrete pile where transporting and in particular driving induce entirely different stresses in the materials than when it is under load as a foundation.

Materials – health, safety and the environment

While it is the properties of a material that may identify it for use in a building, there are reasons why it might not be preferred or even excluded from the ultimate choice. Cost and the availability of local manufacturing techniques have already been introduced, but aspects of health, safety and the environment also have to be considered.

There are many reasons why a material may be hazardous to health particularly during the manufacturing and assembly processes and these will be discussed in more detail in Chapter 13 in the section on health and safety. However, materials have also to be evaluated against their impact on the environment and their contribution to sustainable development.

Society is reassessing its relationship with the natural environment and the demands and stresses we place on it. Much of the concern is associated with our development as a 'technological' society. This new thinking is bound to play a major part in the future in the way we build buildings. Such concerns as energy use, biodiversity, waste and pollution can be directly affected by our choice of building materials.

In assessing a material for its environmental impact it is necessary to trace its entire history, often known as 'cradle to grave' analysis. We can only use the materials provided by nature. Any material that finishes up in a building will have been part of an ecosystem that will be altered by the removal of the resource we require to make and shape our components. In many cases the effect will be local and nature will have the capacity to restore the balance in a short amount of time. However, as the volume of development increases and infrastructure becomes available to exploit resources on a large scale, the restorative powers of nature may be severely tested. The impact becomes more than local to the extent that the exploitation of the earth's resources begins to have global implications.

This is the fear at the beginning of the twenty-first century. Concerns associated with resource depletion, loss of biodiversity and global warming are all affected by decisions to use certain materials in buildings. This is not just in the extraction of materials but also in their transport, for the business of building materials is now part of global trade. There may be a preference for a renewable resource if this can be identified.

Processing and manufacture uses energy and produces waste and pollution. In an attempt to compare the energy used in producing materials we use the idea of embodied energy. It is the sum of the energy used to extract and process materials. It allows not only comparisons between materials but also the evaluation of using recycled material where process energy may be less than that of the original conversion. Waste and pollution also pose a major problem in that returning the by-products of processing to nature may place a burden on the environment it is not capable of dealing with. Disposal strategies are a major concern in evaluating the environmental impact of the choice of materials.

Disposal is of concern not only at the stage of manufacture and assembly but also at the end of the life of the building itself. Specifying materials where the recycling opportunities are limited and the disposal and pollution costs may be high will have to be another part of the analysis of the environmental impact of the choice of construction.

Choosing shape

The shaping of components can bring great benefits in terms of performance. It can give strength by corrugating comparatively thin materials and can provide ornament and finish to satisfy appearance requirements. It has a particular role to play at joints and at points of fixing where shaping such as grooves and rebates can be the major contribution to achieving function. Shaping is, however, dependent on manufacturing processes that require skills, tools and machinery. This adds cost to the component that has to be justified by its utility or a saving in material. It is also dependent on the material chosen; not only for its ability to be worked with current process skills and technology but also for its ability to be worked to fine enough deviations (precision) to achieve the allowable tolerances required for performance.

Early construction methods exhibit the characteristic of minimal working or shaping of the raw materials. Exploiting shape comes with increased resources where skills and manufacturing technique become available. Shaping is achieved either by forming the shape directly from a formless material or by working from simple shapes available from the natural form of the raw material. In some cases these are combined where a crude shape is formed and then refined by removing material to work towards the final form.

Shaping has to be related to an appropriate manufacturing process. Techniques such as casting, rolling or extruding can shape a formless material such as concrete, clay or hot metal. Cutting or machining can then reduce materials to refine or modify the shape. Parts can be added by fixing additional pieces to make complex shapes from simpler forms. Simple forms can also be bent, pressed, beaten or forged into more complex shapes. These processes introduce a range of materials properties associated with workability. They may make a material with all the properties for performance unusable or at least too expensive to use.

Many of these processes are essentially factory based; the range of methods that can be undertaken on site is still limited in most cases by individual skill and tools, leaving pouring, cutting and attaching with some bending as the predominant shaping processes.

Choosing shape and appreciating its performance value and production cost is a major consideration in technological choice. It has particular significance in the issues of deviations and fit to be discussed later in this chapter. Where parts with complex shapes have to fit together there will be particular demands on accuracy that limit the ability of the assembly to accommodate distortions. Hence, while shaping of individual components leads to many opportunities in jointing and fixing this must be considered carefully in terms of the components' relationship to the assembly of which they form a part.

Choosing size

So far it has been established that the level of performance required from the construction leads to the initial choice of material based on a match between the function and a property that can be exploited economically. It has also been established that, for each material, certain processes such as cutting or shaping may be available to make best use of the material and facilitate assemblies that function as whole elements of the building. It is now necessary to establish dimensions sufficient to define the shape, allow manufacture and ensure assembly.

It is helpful to see the establishing of dimensions in three phases working towards a manufacturing or working specification:

- Gross size – based on performance
- Co-ordinated size – based on incorporation with other components
- Working size – based on jointing, fixing, deviations and movement

The gross size will be determined by performance. It represents a minimum size for waterproof materials to keep out the rain or for structural elements to support the loads. It will, therefore, identify the scale and proportion to the solution and may even show at an early stage that a solution is not possible with that material with the chosen shaping in the space available for the construction itself.

(a)   Gross depth based on strength performance

(b)   Co-ordinated depth based on brick courses

(c)   Working depth based on joints and deviations

Figure 4.2 Sizing depth of lintel.

If the gross size appears to be sensible to achieve performance economically and within the space available it is then necessary to look at the dimensions in terms of the detail and how the parts of the construction will go together. This is common, for instance, where the major component is incorporated into brick or block-work. This is illustrated in Figure 4.2 for the sizing of the depth of a concrete lintel over an opening in brickwork. Structural calculations may show that the lintel only has to be 125 mm deep. However, the co-ordinating depth of a brick course is 75 mm (65 mm for the brick and 10 mm for the joint), so the lintel would have to have a co-ordinated size of 150 mm.

Establishing co-ordinating sizes becomes particularly significant in more industrialised and prefabricated buildings that are often designed within a dimensional discipline around a module, leading to a basic or modular size being established. The most common module is likely to be 300 mm. When selecting factory-made components available from manufacturers' catalogues, the economics of production means that they are often only available in a limited number of sizes. Although these will be based on a module most likely to maximise their potential for integration with other components, they do limit the choice without incurring additional cost.

The co-ordinating size does not take into account:

- The need for space for joints and fixings
- Allowance for deviations and movement

These have to be considered together, as in many cases it is the joints that will have to allow for the deviations and movement and the fixings will have to accommodate both deviations and movements if the assembly is to fit together on site and maintain its integrity in operation. Consideration of these aspects will lead to a working or manufacturing size.

Joints are a necessary part of the assembly process to ensure fit in that they can take up deviations in components. Fixings are required to hold assemblies together and these also take up space. Sometimes the joint is also the fixing. The material of the joint and the fixings dictates their own dimensions (with its own deviations). If the fixing is in addition to the joint and this cannot be accommodated completely within the component or in the joint then space has to be provided for the fixing itself. This will have to include any access required to complete the fixing on site. In some cases thought has to be given to space necessary to remove the fixing for repair or maintenance.

In the example of the lintel the joints in the brickwork are 10 mm. This will be a mortar joint that will allow for deviations in the brick. It is now possible to establish the working size for the lintel at 140 mm. For a full specification of the depth it is necessary to be clear that in the manufacture of the lintel any deviation from the 140 mm will have to be limited as these also have to be accommodated in the 10 mm mortar joint. Figure 4.2 suggests an allowable deviation of ± 2 mm. It is now necessary to ensure that the material can be manufactured or worked within these deviations or there is a risk that the lintel will not fit with the brickwork.

Working sizes also have to take into account any allowance for dimensional changes that may take place during the life of the component, known as movements or induced deviations, and these are discussed in more detail in Chapter 12.

Dimensional deviation and fit

When establishing a working size and thinking about how assemblies might fit together it is necessary to think about three things:

- The size of the space into which the component has to fit
- The size of the component and its position in the space
- The gap that will be formed between components for the joint/fixings

Generally a component can be said to fit if it can be positioned within the space so that the gaps allow for effective fixing and jointing. During construction, all of these, the space, the component and its position and the gap, will vary: they will all need to have tolerances established.

In what is often thought of as traditional construction, trade processes control fit. Procedures such as 'building in' and 'cut to fit' allow for dimensions to be set by the construction itself as the work proceeds. It requires judgment as to how much reduction should be made for fixings, jointing materials, gaps or movements, but these are part of the craft tradition. It is not a question of analysis but one of common practice. One person can take all the deviations into account at one point in the assembly procedure in one judgment.

With the use of components and the development of prefabricated construction, building in and cutting to fit is no longer an option. There has to be some understanding of these deviations as a basis for analysis to decide on final working sizes before manufacture. With elements of the construction being formed from many components this may have to be based on a statistical analysis of the deviations that can be achieved in manufacture and assembly and will be closely associated with the detailing of the joint.

Consideration of the production process will soon reveal an inability to reproduce the specified dimension for the space, component or the joint in practice without deviation. This is not a reference to people making mistakes, the detection of which is a management problem, but an acknowledgment that materials shaped with the most skilled hands, even assisted by machines, will produce components and assemblies that will deviate from the specified figures, and this is a technical problem. These deviations are 'induced' by the chosen production processes. They will be within limits dependent upon the type of material, the level of skill and assistance provided by tools and machines plus the conditions under which the individual has to work. It is possible to identify these induced deviations or characteristic accuracies from a particular production process.

The particular dimensions that are actually achieved for the space, the components and the joint will vary within a range of deviations, most of which will be close to the target value with a decreasing probability of larger deviations being produced. This probability follows a pattern

(a) Process producing limited deviations

Normal distribution
of outcomes

Target value

(b) Process producing wider deviations

Normal distribution
of outcomes

Target value

Figure 4.3   Identification of achievable deviations.

known as a normal distribution, as shown in Figure 4.3. This tells us that you cannot predict the size of the space, component or gap in advance, only the probability that it will be within certain limits. This deviation from the specified value is also true for the specification of the quality of the material.

Because these production deviations are a reality it is necessary when evaluating the dimensions of components and assemblies to consider how such deviations will affect the performance. Some limits will have to be placed on these induced deviations for, if too great, the performance will be impaired. These limits, or allowable deviations, are known as tolerances, or what that particular assembly can tolerate before performance may be affected.

Tolerances are quoted as a range within which an assembly will successfully perform. They are limits where it is deemed that components installed outside these limits will fail, or at least risk failure in the assembly. They are represented graphically in Figure 4.4. They take the form of a ± value from the specified size. While most specified sizes are set as an average where equal deviations on either side can be tolerated, some specifications are maximums (or minimums) allowing deviations in only one direction.

Knowing the probability distribution for the chosen production process and having an allowable tolerance set by the detailing allows an estimation of the probability of fit. If the allowable tolerance is greater than the deviation that will be achieved for the great majority of the components from the chosen production method, then the probability of assemblies failing will be small.

It would be usual to ensure that the detailing chosen can tolerate the deviations inherent in current production processes. However, in some designs it may be necessary to devise revised or even new production processes that can achieve the required tolerances, but the cost of this development and the risks involved must be appreciated before the proposal is adopted.

It is true to say that before a solution can be specified with complete confidence, an analysis of both allowable tolerances and inherent deviations (characteristic accuracies) must be made. This will involve an understanding of both the behaviour of the construction under working conditions and the outcomes of the production process before a solution can be said to be a viable proposition against performance criteria.

(a) Specification as average

Specified value x

t   t

allowable deviation ± t

(b) Specification as maximum

Specified value x

t

allowable deviation +0−t

Figure 4.4   Specification of allowable deviations.

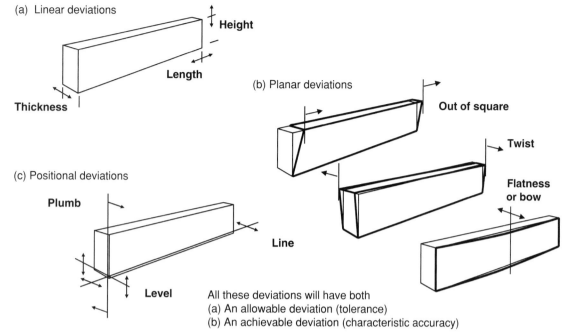

Figure 4.5 Possible dimensional deviations.

Types of dimensional deviations

The previous section looked at the relationship between inherent deviations and the tolerances that can be allowed to ensure fit and operational performance. In defining the component, its space and the gaps that will arise for the joints and fixings, it is important to realise the ways in which deviations can distort as well as just affect size and position.

Figure 4.5 shows a simple rectangular solid component, but this could also be the space available for the component or the space representing the gap between components for the joints and fixings.

Variations in size can be seen as linear deviations affecting the three principal dimensions. However, these may not be the same at all points across the component and this will cause distortions or planar deviations of square, twist or bow. The third set of deviations is positional associated with the line, level and plumb aspects of site control that ensure the spatial relationships are maintained.

Spatial relationships

There are two ways of defining relationships that can be employed to specify position:

- Referenced – specifying a position in space for each individual part within a common reference framework
- Relative – maintaining a positional relationship with a number of other parts of the construction

The use of each will depend on the scale of the building and the stage of construction. Since the process of construction is predominantly building one part after another, dimensional deviations can become cumulative if relative positions alone are used. It may be that fit in parts of the building may become difficult to achieve within their own allowable deviations because previous specifications did not take into account the possibility of 'knock-on' effects. Components can move further and further away from the intended spatial arrangement that the relative dimensions were intended to maintain. This is

(a) Structural grid on plan

Initial setting out to
relative to co-ordinates
or other existing features

Co-ordinating grid dimensions

(b) Floor heights in section

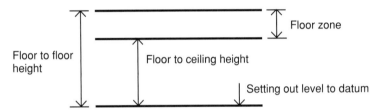

Floor zone

Floor to floor
height

Floor to ceiling height

Setting out level to datum

Figure 4.6   Dimensional control
framework.

despite the fact that all previous dimensions have remained within specified deviations.

For construction where 'cut to fit' and 'building in' can be employed, spatial relationships can be controlled to maintain the detailing of the parts and the joints between them. By working within tolerances, parts can be adjusted for the specific circumstances. In this way the all-important joints and fixings can be produced in the required position and within the acceptable limits of their allowable deviations. In modern construction the process is more the assembly of manufactured components of predetermined size. The preformed parts tend to be built one on top of each other. The possibility for dimensional creep or moving outside spatial constraints may be high if too much of the specification is relative.

It is, therefore, common to specify most of the early stages of the construction to a reference framework. Sufficient of the construction has to be specified and controlled in this way until enough 'real' construction exists so only a few

parts have to be added to complete that section of the work.

The frameworks developed to control position in this way are usually grid lines on plan and finished floor levels on elevation. These are illustrated in Figure 4.6. In both these cases the 'lines' as they appear on plans and sections are actually planes that 'box up' space and from which positions can be specified.

The control 'box' should not be too small because the positioning of finishing, fixtures and fittings can be relative to the major construction so long as they have been well controlled by referenced dimensions. For this reason the grid system chosen is usually associated with the structural layout for the building. Every effort would be made to keep it simple with repeating dimensions to avoid mistakes in both design and production phases.

It is these grids and floor levels that would be the basis of site dimensional control. For all but simple building this would usually involve the use of surveying equipment, tapes, levels and theodolites. Advances in this type of equipment including electronic distance measurements particularly associated with theodolites in total station equipment have led to the use of co-ordinates for site control.

Joints and fixings

Where two or more components or elements of a building meet there is a need to consider how they should be joined. Where one or more of these parts requires support or restraint from the others there is a need to consider how they should be fixed. Joints and fixings are the all-important details that:

- Hold the construction together
- Allow it to be assembled
- Ensure the performance of the fully erected building

Like components, they are made of materials, and have shape and size. They have to work within gaps or their own space between components. They only work within certain limits of that size (allowable tolerances) and can only be produced within limited accuracies (induced deviations). They then have to be positioned within specific limits. They are, in fact, subject to the same considerations of performance and production criteria as the components they join and fix together.

Joints and fixings arise to facilitate the creation of the whole from a series of parts. They are, therefore, crucial in the overall behaviour of the building. They influence its stability, its ability to accommodate movement and provide the continuity of performance of the elements of the building of which they form a part.

They also play a major part in facilitating the assembly process, particularly where many of the components are manufactured off site, when the joints and fixings become the focus of the site activity. Buildability, safe working and skill requirements become a matter of handling the components and the completion of the joints and fixings.

In common detailing it is often difficult to separate out what is a joint and what is a fixing. Fixing implies a support or restraint function with load transmission capabilities. Joints tend to describe the whole detail where materials, components or elements come together. The joint may have many functions such as weathering, fire, sound or movement and may not include a structural requirement that implies the action of a fixing. In some cases components may be fixed independently and are stable but need to be jointed to achieve continuity of function across similar components. The particular requirement of a joint to isolate movement in a building needs special attention. Fixings may also have to be chosen to allow movement in at least one direction or the joints may have to be created totally independently of the fixings.

A number of joints and fixings are shown in Figure 4.7. In traditional timber jointing, such as the mortise and tenon, the detail is entirely based on shaping the components to be joined, which then becomes the fixing of one part to the next. Although glue may have been used, often no other components were necessary. The joining of plaster to a wall also needs to provide fixity. With old plaster systems, laths allowed the wet plaster to form a shape, giving a mechanical key, while modern plasters can achieve fixity with adhesion alone. The jointing of plasterboard to studs is achieved with a mechanical fixing, a nail or screw.

Joints are often subject to performance requirements other than fixing, such as appearance, weatherproofing or fire. The joint to be made between the sheets of plasterboard in the stud system shown in Figure 4.7 needs to take into account movement that might cause cracking of the jointing material, spoiling the appearance. Joints have to be reinforced with tape to resist these forces. Cladding panels on the external face of a building are normally fixed to the structural frame independently, being supported and restrained with a range of fixings such as bolts and angles. The joints between panels then function to allow for deviation, both inherent and induced, and provide continuity of appearance, weathering and other functions of the cladding system itself.

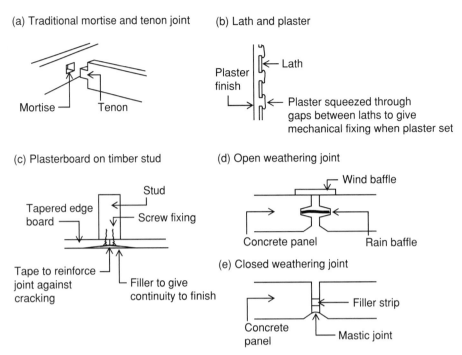

Figure 4.7 Joints and fixings.

Joints between structural members have a particular significance since the behaviour of the joint determines the distribution of stress in the structural system as a whole. The detailing determines whether the joint can rotate or is stiff (fixed). This determines the way the forces are transmitted from one element to another. More detailed analysis of this aspect of jointing and continuity between members is provided in Chapter 12.

The function of joints and fixings to facilitate assembly needs to consider the issue of deviations. They have to be detailed to work within acceptable deviations, but they can also be used to absorb or even out deviations to maintain the position of components. Joints and fixings that can maintain local allowable deviations ensure the achievement of fit throughout the building.

There is now a wide range of fixings, the exact form being determined by the nature of the forces being transmitted, the part to be fixed and the structure to which it is being attached. Its form will also be determined by the need for accuracy and adjustment in positioning of the part to be fixed. Options include welding, gluing, nailing, screwing and bolting and may involve brackets or plates. Specifications must include shape, size and materials as well as the pattern or position of fixings. The choice is also influenced by economics, which can be significantly affected by the tools required and the influence of 'deskilling' associated with the overall construction specified.

Performance and fit depend on the careful consideration of the joints and fixings. Problems are created and resolved around such details. Careful choice and specification of the materials and components has to be matched by a similar attention to detail on joints and fixings if the building is to work as a whole.

Summary

(1) In making a choice for the construction it is necessary to specify the materials, shape and size, and spatial relationship of the parts.

(2) Specifying materials starts with identifying appropriate properties for performance. This then has to be linked to the availability of the material in an appropriate form with reliable supply within acceptable deviations. Health, safety and environmental considerations are also significant in making the choice.

(3) The ability to shape materials brings great benefits in using the properties of the materials efficiently in stable configurations, particularly at joints and fixings. Shape is particularly important to appearance and in the fine detail that ensures performance. Shaping, however, requires tools, machinery and skills that must be available at reasonable cost.

(4) Size has to be specified to define not only overall dimensions but also the required shapes. The gross size is identified by the basic performance requirements while the co-ordinating size is determined by the need for the parts to work together. Given the need to join and fit parts together, the co-ordinating size will have to be modified to a working size that will then need to have deviations identified to ensure the detail can be assembled and will perform.

(5) Each detail of construction will have to be evaluated for its ability to be assembled within these deviations and still perform. It is not possible to produce perfect construction, so it is necessary to establish what deviation could be allowed or tolerated without loss of performance.

(6) Specifying and controlling spatial relationships can be to a common reference system or by positions relative to other parts of the construction. Given the existence of deviations, if a reference system is used there will be less chance of cumulative deviations, reducing the probability of failure due to dimensional creep.

(7) Joints and fixings allow the parts to act as a whole and the assembly process to be achieved. They influence stability, the accommodation of movement and provide continuity of performance in the elements of the building as well as the building as a whole.

(8) Fixings imply a load transfer function while jointing involves many functions which may, or may not, include a structural role.

(9) In facilitating the process of assembly, jointing and fixings have to be chosen that can accommodate deviations in order to maintain spatial relationships and the performance of the building as a whole.

5 Defining Conditions

This chapter identifies the fundamental importance of the creation of both physical and social conditions to the choice of construction. It establishes the need to specify internal requirements, define external conditions and then to identify the contribution of the activities of the occupants and the construction itself to the internal conditions. These conditions will be ever changing and therefore the whole system must be seen as dynamic; a series of flows seeking equilibrium. The chapter concludes by identifying how this process can be applied to the choice of internal elements such as floors and partitions, and introduces the idea of microclimates within each element of the construction itself.

Physical and social conditions

It can be argued that buildings are fundamentally concerned with providing physical construction to create the conditions for the well-being of those who use the building. This statement would raise in most people's mind the modification of the natural environment: from the weather conditions outside the building, to a less variable and more comfortable and safe environment inside. This is clearly a primary technical role for the construction, and therefore a clear understanding of the external prevailing conditions and a clear specification of the required internal conditions are of primary importance before any process of choice can begin.

Increasingly the external environment is being significantly modified by human activity with noise and air pollution so any understanding of the external environment must be more than just the natural weather conditions. It becomes necessary to understand the impact that producing and operating a building will have on the local, regional and even global environment.

The case for the need to understand the interaction of physical conditions is easily understood, but solutions chosen only against criteria of modification of the physical environment are not sufficient. Buildings are created in a social context in which they serve social functions that influence performance requirements. This simple idea is shown in Figure 5.1 which illustrates that the user's interests in modification are concerning both the physical and the social worlds. If an activity can be carried out effectively within the limits of prevailing external conditions, and it is socially acceptable to do so, there will be no requirement for a building in any form.

It is important to remember that buildings have an impact on social conditions with both the internal and external spaces they create. While an understanding of social behaviour is beyond the scope of this text it is well established that building may be part of social problems but can also be part of solutions. This is recognised in many aspects of urban design in areas such as crime and community safety.

The social context also determines the resources that become available for building. There is a need to understand the influence of the economic, political and cultural environments on the choice of construction set against its cost and its value in its wider social interactions. This has an influence in the specification of the required internal environments. The expectations of the user and the variations in internal environment deemed acceptable are set against these wider social standards and the economic possibility determined by the amount of income clients are prepared to spend on their buildings.

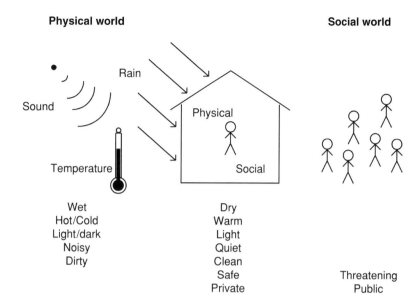

Figure 5.1 Physical and social environments.

Activity and construction-modifying environments

It is clear that construction modifies the external environment to create the desired internal conditions. However, the activities within the building will themselves contribute to the internal environmental conditions. Cooking generates moisture, heat and smells, machinery generates heat and noise, and cleaning activities generate waste. They will in turn, if detrimental to well-being, have to be controlled, which may make further calls on the technical demands of the construction.

It also has to be appreciated that the construction chosen may itself modify the environment. This is particularly true of the services in building that often create heat, fumes and noise, but it can also be true of the passive construction. Poorly detailed insulation to resist the passage of heat to provide the thermal environment with lower energy inputs can induce condensation and create damp conditions. Finishing materials can give off toxic fumes in fire. Both are examples of the choice of construction affecting the achievement of the desired internal conditions.

The dynamics of the system

Although a stable internal environment may be the objective, it is clear that the external conditions and the levels of activity within the building will change in a variety of ways. This will produce constantly changing conditions. The whole system can be seen as a series of flows seeking equilibrium, as shown in Figure 5.2. This dynamic makes the process of control of the internal conditions a significant aspect in the process of choice.

Predicting the possibility of exceeding design limits in dynamic systems needs an understanding of both the variability of the conditions and the response of the construction providing the modifying effect. Indeed, a major consideration in choosing a solution may well be its ability to respond in sympathy with the expected magnitude and speed of changes in the environment. This has been recognised, for example, in the advantages of high thermal mass for building in limiting the use of energy for cooling on hot days to limit the variations in the internal temperature between day and night.

To enter into an analysis of the behaviour of the construction it is necessary to define the following:

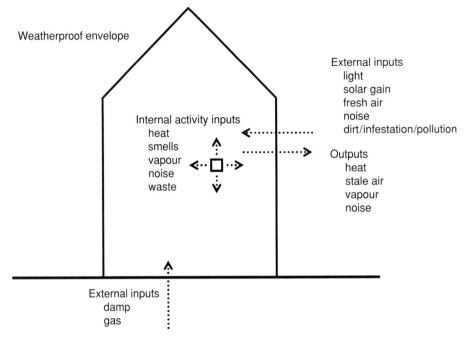

Figure 5.2 Environmental flows – a dynamic system.

- Target internal conditions and acceptable limits within which the activity can operate
- Design limits in the external environment and their pattern and variations
- Inputs to the internal environment from the activity and construction including the response to changes in conditions

Target internal conditions

Defining the target or ideal internal conditions is a question of physiology and psychology in the case of people, engineering in the case of machines and spoilage processes in the case of stored material. Aspects of the environment important to the well-being of life, machines and materials have much in common. Temperature and dryness are often the most significant to the well-being of all three, although ideals and limits may be different. In cold stores the temperature is clearly dictated by the stored food. Any people or machines that also have to operate in the store have to be protected by other technologies, e.g. clothes in the case of people. In addition to temperature and dryness, consideration has to be given to aspects such as humidity, light, noise and air quality. The recent concerns over air quality, associated with hitherto unrecognised contaminants which occur in very low concentration such as radon, illustrate how new aspects of environments can come into play as we strive for more demanding and less variable conditions.

People, machines and materials do not, however, need precise environmental conditions and can operate within limits. Indeed, in the case of people, constant, unvarying conditions can cause fatigue and some variations are desirable. Specification of internal environments is greatly influenced by social expectations and the resources clients are prepared to devote to the building. With low levels of resources available for the construction, wider limits of environmental conditions may have to be accepted. In the early factories of the industrial revolution, concentration on machine efficiency often made noise and dust levels far from ideal for the workforce.

Social expectation is often now embodied in law to ensure clients make resources available to maintain healthy and safe environments within their buildings. Many regulations and advisory standards now set design limits for comfortable, safe and healthy environments.

Establishing external design conditions

Establishing external conditions, both climatic and social, is clearly a different task from setting internal conditions. It is necessary to understand the broad patterns of events, with the probability of extreme conditions being identified. The broadest climatic patterns can be identified by global classifications. The fact that the UK lies in a temperate zone gives an overall picture of conditions. However, regional and even local conditions may well determine the pattern of variations and extremes. The South West has the fewest frosts and the highest winds occur along the west coast of Scotland. Variations also exist between urban and rural areas where the pollution and the modifying influence of existing developments can have a significant effect, an example of which would be traffic noise. All these may influence the orientation of the building on the site and the choice of materials and detailing to ensure the required performance is achieved. This would also be true of the social conditions, indicated by broad patterns such as inner city, suburban or rural classifications. These may give an indication of the social conditions, but often much more local knowledge will be required if the performance of the construction is to be achieved in practice.

The external conditions need to be understood in order to evaluate most aspects of performance. The need to define climate for environmental control is straightforward; temperature variations and sun paths for heating and cooling requirements, rain and wind conditions for weathering to ensure the building remains dry. However, external conditions also need to be known for structural performance and for durability. Both snow and wind place loads on the building structure but in different ways: sunlight fades colours and embrittles plastics; saturated bricks subjected to frost may disrupt under the internal pressures created during the growth of the ice crystals.

It will be noticed from the above examples that combinations of climatic factors are often the worst design conditions. Wind and rain can be the most testing conditions for water penetration; snow and wind may give the worst loading conditions; wet weather followed by frost causes damage to brickwork. Defining single factors may not reveal the most testing conditions. The pattern and variations in the climate have to be understood well enough to visualise the significant combinations as well as individual extremes.

Establishing variations and extremes requires knowledge of conditions beyond that of the daily cycle or the annual seasons. Extremes do not occur every year but at longer intervals. Some of these seem to be identified with long-term cycles or trends and some seem more random. Since most buildings have design lives measured in decades then cycles, trends and random extremes that happen over these time scales need to be investigated.

For each stage of the analysis (environmental, structural and durability) it is necessary to visualise the most testing set of conditions to maintain performance. For durability this will be dependent on the materials chosen, for the agents of decay are specific to each material. Each set of conditions will then need to be defined in the most appropriate way. Knowing annual rainfall gives a broad indication of the magnitude of the problem faced in weathering a building and providing drainage but will not be the whole picture. Occurrence during the year and some measures of intensity are required. Rain that occurs in a single season of the year at high intensity poses different problems to rain that occurs in light showers throughout the year. Intensity over a period measured in minutes will determine the size of drainage system, while lower intensity over some hours, particularly if associated with high winds, may determine the detailing of weathering to the walls and roof. In another set of circumstances, prolonged light rain may penetrate porous materials sufficiently to cause dampness on the inside of construction well able to resist a sudden downpour.

While this rainfall example is directly related to the physical climate, the principle of visualising the most testing conditions is true for other performance requirements. Changes in social environments, say with crime and terrorism, also need to be established in some way that allows a design solution to maintain performance of the internal environment even with the perceived levels of threat. Crime figures may allow some quantitative analysis of probability, timing (day or night) and mode of entry which allows choices to be made as to how best to maintain performance.

Since the external conditions are variable, it becomes a design decision as to which limits are to be taken into account. Generally, design involves establishing the probability of the occurrence and a level of 'value' on the disruption, danger or cost of the consequence of the building failing to perform. Rare but devastating or life-threatening events may be given similar importance to common occurrences that cause inconvenience or a lowering of operational efficiency. It requires the client to assign value to loss of performance associated with the prevailing social and economic conditions. It may also be subject to the law for in many cases loss of performance poses a danger that the client may not value as highly as society in general. Over the years this has been deemed to be the case with loss of life in a fire and this has led to extensive regulations.

Yet another aspect of the analysis of proposed construction that will require knowledge of the external environment is its manufacture and assembly. Conditions influence the accuracy that can be achieved, leading to variations in the achievable deviations. Protection to both operatives and the partly completed work will be influenced by the condition. Efficiency and hence cost will change with conditions. The way the conditions are defined will again vary depending on the purpose of the analysis. Number of hours of daylight or frost days per month may now be the most appropriate way to seek the information about the external environment in order to identify the effects on the decision to specify works to be completed on site in the winter.

Social conditions also influence decisions on the production process. The labour market, local,

regional, national and even international, has a major influence on the availability of skill and the cost of labour. Site security will be influenced not only by local crime rates but also by legal and social obligations to keep a safe site that protects both operatives and the general public.

Inputs from the activity

For each aspect of the activity, the types of outputs and the quantity of each will have to be established. People within a building will produce varying amounts of heat, moisture vapour, other products of exhalation, odour, etc., depending on the activity being carried out. While some of these may be helpful in, for example, heat gains in winter to maintaining temperature, others will militate against clean air. In another example, the contribution of moisture vapour can have a significant influence on the formation of condensation if ventilation rates are not sufficient. Machines can produce noise and heat and there are concerns over the electromagnetic fields emitted from electronic equipment. While most stored material is inert and contributes nothing to the internal environment, this may not be true if the material stored is organic. Bulk storage of straw can, under some conditions, generate sufficient internal heat to catch fire.

As with the external environment, broad patterns of these internal contributions may be established, but questions of variations in duration and rates of activity will have to be determined. Designers often look for the single figure to represent inputs from the activity, but if variations are great or change rapidly it may be necessary to analyse the dynamics of the change to match the response of the building more carefully. If data are not available for particular activities, studies will have to be carried out to determine the rate of net inputs into the environment for appropriate construction solutions to be developed.

Inputs from the technical solution

Clearly the effects of the construction and the services cannot be defined before the construction and services are chosen and must, therefore, become part of the process of choice itself. Problems associated with the building in this respect have to be foreseen and minimised at the beginning of the process by the broad strategies developed rather than by resolving problems at the end in the specification and detailing.

Deciding on broad strategies involves decisions on all the performance requirements and is therefore a careful balancing act. This includes the balance between passive construction and the active services. With increased demands for quality in the internal environment with smaller variations, active services are difficult to avoid. Unfortunately it is the services that are most likely to generate outputs that affect the internal conditions and use energy. Some input can be used to advantage. Heat recovery from lighting can reduce the size of the heating system and reduce overall energy use. However, the noise from the fans necessary to recirculate that air may be transmitted through the ducts. This in turn will need decisions on the use of silencers or sound absorbers in the ducts. The materials for these now have to be chosen with care as fibres from the sound absorption material may be picked up in the air stream and carried to the rooms the ducting system serves. Designs of building that can take advantage of natural ventilation generated by wind and other pressure differences have considerable advantages, but this has to be part of the overall design concept and not a question of detailing.

Most construction materials are inert and therefore add nothing to environmental conditions. Moisture from wet construction at the early stages of occupation of the building has been largely overcome with the use of dry forms of construction.

Choice of interior elements of the building

So far the analysis of internal and external conditions has focused on the inside and outside of the building. However, buildings are rarely used for a single activity. Although there will be an overall purpose for the building, fulfilling that purpose for the user will almost certainly involve a series of interrelated but separate activities. In specifying internal elements, floors and partitions, in the building the notion of an 'external environment' may refer to another internal space used for a separate activity. In this case the internal/external labels are interchangeable.

Problems posed for the construction to solve will depend upon which way the undesirable elements of one environment will affect the second. The task of the construction depends upon which way the flow disturbing the desired conditions will take place. It may become the task of an internal partition to modify sound transmission in one direction and heat in the other if it is between an unheated factory area and an office.

Inside the construction itself

Although apparently not of immediate interest to the user, the process of modifications of environments across construction creates 'intermediate' or 'micro' climates within the construction itself. They can become highly significant to the running cost or life expectancy of the construction and that is clearly the concern of the building user or operator. Perhaps the best example of this is the formation of interstitial condensation causing damp when it is apparently dry on both sides of the wall or roof. The wetting of insulation will immediately reduce its effectiveness and, if persistent, may create the condition for rot or corrosion, leading to premature failure.

The analysis of these intermediate conditions that rarely reach steady state is another task for the technologist seeking to provide satisfactory, continuing performance throughout the range of environmental conditions that may prevail.

Summary

(1) Building is concerned with providing physical construction to create environmental conditions within the building that are modifications of those prevailing externally. These modifications are required to fulfil both physical and social functions.

(2) In providing physical construction, the building will have an impact on both the physical and social environments and these must be understood and be part of the analysis.

(3) Both a target value and an acceptable range have to be identified for a number of parameters to define the internal conditions to ensure a comfortable, healthy and safe environment. These will depend on the activities being carried out in the building.

(4) The activities within the building generate inputs that will modify the internal conditions. If these inputs take these conditions outside the desired range, the building will have to provide the necessary control to maintain the required internal environment.

(5) Under operational conditions the interplay between changing external conditions and inputs from the activities and the construction itself must be seen as a dynamic system: a series of flows seeking equilibrium.

(6) In addition to understanding how internal conditions are to be maintained it is important to understand how the choice of solutions will impact on the environment in general.

(7) Defining the external conditions, both climatic and man-made, is necessary not only for the environmental modification role of the construction but also because it loads the structure and deteriorates the materials of which the construction is made.

(8) Because external conditions influence so many aspects of the analysis of a proposed solution, the same aspect of climate, say rainfall, may have to be defined in a number of ways depending on which part of the analysis of a solution is being carried out. Aspects of climate are also important when deciding on appropriate methods of manufacture and assembly.

(9) Internal elements of construction often separate areas of different activity requiring different internal conditions. In this case the definition of the 'external' environment is another but different set of internal conditions.

(10) When modifying environments the construction has different conditions on either side which means that the conditions within the construction are neither internal nor external. Identifying what the micro-environment is within the construction and its rate of response to change will have to be analysed.

6 The Resource Base

This chapter introduces the four key resources required for the successful creation of buildings. Materials are identified as the primary resource being associated directly with the construction of the final building. The second is the knowledge and skill held by people undertaking design and production in order to exploit the properties of the materials. Machines enable the application of materials in forms, sizes and shapes not possible with human effort alone. Finally, money as a resource has to be deployed by clients and the players in the construction industry.

Introduction

Resources are normally associated with the activities of management, organisation and control. They have often been quoted as materials, people, machines and money that have to be employed in the construction process. In management terms these four resources represent the inputs for the construction of the building which need to be obtained, deployed and controlled. For the technical choice of the construction the understanding has to be different to that of management, but the idea of four basic resources is a good starting point to think about the evaluation of a suggested construction solution.

A knowledge and understanding of resources becomes central to the analysis of buildability, cost and social interactions in the evaluation of the suggested solution. The whole concept of buildability is centred on the capacity to successfully obtain resources to achieve the required specification that will realise performance. Costs arise only from the deployment and control of resources, the price of which are determined by the prevailing economic and political conditions. Society has an interest in the impacts on the physical and social environments of the use of resource for the process of construction as well as the newly created resource of the building itself when in occupation.

Materials – the primary resource

Since materials are one of the basic variables of the final construction they can be seen as the primary resource. They are identified as a potential for the construction because of their properties to achieve performance, as discussed in Chapter 4. However, there is also a need to understand our ability to bring the raw material to a finished state suitable for incorporation into a final assembly. It is bringing the raw material to a finished state, followed by fixing in the final position, that determines all other resources required.

The use of a material is, therefore, dependent not only on its potential properties but also on the availability of resources to process, manufacture and assemble the building. It is the existence of (or at least the economic and political will to release) the resources to bring the material to its finished state that determines the success of a chosen construction form or detail.

In societies where resources are not readily available, construction is often limited to relatively unprocessed materials of local origin. The properties of the material may be difficult to predict, with variations in the quality of the material brought to the site. This means either a process of selection giving a high level of waste (often economically unacceptable) or the design has to assume only limited performance from the material, leading to larger sections. Many of the developments in processing materials seek to limit the variation as well as enhancing the basic natural properties.

The consequences of an increased manufacturing base and infrastructure are not only a guaranteed quality but also a greater range of materials. The existence of these new materials does not, however, lead directly to an ability to incorporate them into a design with the expectation of successful construction. There is always a need to be able to detail with the confidence that the required skill to work the materials for assembly into the building is available.

In addition to understanding the properties of a material for performance, its workability properties have to be known. Very few materials are useable in the form in which they are found in nature. There are whole industries concerned with the processing of materials upon which buildings rely for the availability of components. In some cases, such as timber, this is mainly careful conversion so the potential materials properties are maintained. In other cases, such as plastics, the formulation and processing can determine a wide range of properties, although only within the basic behaviour established ultimately by the long chain molecular structure of this group of material. In both the cases of timber and plastics, not all types have similar workability. Straightness and closeness of grain for timber and the basic distinction between thermoplastic and thermosetting plastics, for instance, influence how they can be used in practice.

The ease with which it is possible to perform operations on materials is key to choosing size and shape for components as well as possible fixing methods to provide the assembly. The ease with which each material can be cut, bent, poured, extruded, etc. without loss of its useful properties determines the manufacturing and assembly methods. This in turn determines the economics of the construction.

This may lead to a range of materials being used for similar components. Bricks, although usually associated with clay that is moulded or extruded when wet and then fired, can also be made of concrete, which relies on chemical reactions that take place at ambient temperatures. While this gives a product with many similar properties, some are different. Clay bricks expand in early life, while concrete bricks shrink; clay bricks can be dimensionally and visually irregular, while concrete ones have fewer deviations. These are endowed by the properties of the material; they are not a basic attribute of the component.

The materials we use are not limitless and their conversion and processing have an impact on both the built and the natural environments. As discussed in Chapter 4, this has perhaps been the major realisation of the late twentieth century and its influence on the choice of materials for building is likely to be profound. Concern for the impact of extraction, depletion rates, energy for conversion, pollution and waste and ultimate disposal leads to cradle-to-grave analysis, showing the true implications for the use of each material on the environment. Materials with all the right properties and workable with the skills available may not be chosen because of their environmental impact.

Knowledge and skill – the human resource

Many animals can be identified as builders. A few higher primates seem to have the rudiments of using tools. Humans alone seem to have the abilities to take this to the level of complexity we observe in the world today. This intellectual power when applied to the development of the skills to design and build is the second great resource we have to create buildings.

It is the development of knowledge and skill that determines not only the genesis of ideas about how we might build but also our confidence to try. Human activity is predominantly social and increased social complexity demands a greater range and purpose for its buildings. Given a sympathetic cultural environment, increased economic activity releases more resources to technological activity, which will be reflected in the methods used for building. Perhaps more significantly, the cultural environment determines the speed at which progress in technological development takes place. If developments move too far ahead of knowledge and skill there is an increased risk of failure.

Every time a new idea is put into practice it is a trial that may fail due to an error in our understanding of the material or the behaviour of the construction. Technology derives a great deal of its knowledge and know-how from practice as well as the sciences. It is an assessment of this total knowledge and skill that has to be available before the risk inherent in the choice of construction can be evaluated. In a complex social environment this knowledge and skill is unlikely to be available in any one person or even a small group of people. Communication between those with different aspects of technical understanding becomes increasingly important.

This resource, characterised as knowledge and skill, is identified in social organisations involving people. The operatives, designers and managers are most directly involved, gaining their knowledge from their education and training and then from their daily experience. Material and component manufacturers also generate and accumulate information and knowledge that is vital to the process of incorporation into the building. Behind all this lies the research and development activity often not directly related to the construction industry. This feeds in knowledge that can be called upon by practice to improve the chance of sound decisions and lower the risk of failure.

This knowledge is fundamentally, but not exclusively, technical. Design is a key ingredient in the choice of construction to provide fine buildings but is of limited value without an understanding of the technical issues involved. Organisational and motivational skills are vital to the manager but can be misplaced and ill-directed if the technical issues are not well understood. The skill of the operative to use tools cannot be isolated from an understanding of technical matters, particularly where new construction is being introduced. The judgements being made by the operative on, say, deviations (tolerances) are not just about the knowledge of the workability of their material but may have consequences for subsequent operations of assembly or the long-term performance of the building.

It is the capacity of each individual to make technical judgements combined with their individual skill that is the key to identifying the availability of appropriate resources. It is the technical understanding that has to be shared and communicated so that each individual may make sound decisions in order to limit the risk in the overall construction process.

Plant and machinery – the technological resource

With the right people with the knowledge and skill to select, process and work the materials into effective components and assemblies, it is possible to produce buildings. These buildings would, however, be made of a small number of materials and would be limited in size. While hand tools will be associated with the operative it is the use of plant and machinery that expands the technological possibilities.

Although often seen today as the way to greater productivity and economy, the machine was initially an enabling resource. It allowed solutions to be chosen that could only be built if the machine itself could be conceived and constructed. It formed a chain of technologies that had to be available before a particular construction could be realised.

In the early days of large construction, such as cathedrals and early civil engineering works, these machines would have been designed and built by those responsible for the construction of the works themselves. They were often associated with lifting or simple materials processing. Manual skill would still be employed to shape and finish most of the components and fix them into their final position.

Although machines are still fundamentally about breaking away from the limitations of the strength of an individual with only hand tools, machines may now make operations economic and therefore become the preferred choice as they lower cost or increase quality.

This process of mechanisation is not just associated with the site assembly process. Behind the expanding choice of materials lie developments in processing plant. Behind the expanding choice of components lie mechanised factory processes. The technological chain that allows choice of construction for the building itself is now extensive.

Machinery has long been available for processing and lifting materials but often by mechanical advantage still using human power. Developments have been greatly influenced by alternative sources of a power supply. Individual power units can now be devised for very large and very small machines, making the availability and versatility of plant and equipment very great. However, most machines still require human knowledge and skill to guide and operate them. Perhaps the most radical change in the relationship between the choice of construction and machinery will come with the introduction of machines that take over the operator functions. With the application of microprocessor technology to automation of construction equipment, the conventions of the material/labour relationship may be supplanted by material/machine protocols. This can be seen in the use of computer numeric control (CNC), now available for structural steel fabrication, which can take its information directly from computer-held design information.

The increasing use of power often involves the burning of fossil fuels and this will have an environmental impact that needs to be taken into account in the choice of construction.

Money – the enabling resource

Money is fundamentally different from the other three resources in that it is not used directly in the construction of the building. It is not technical in nature and is therefore the most easily transferred to other areas of economic activity. Its use for building has to be seen as 'good value' by those who are willing to deploy it. In the past, patronage for great building has been directed towards status and the association with fine things. More typically today it is about return on capital and this determines much of the ethos in which construction is undertaken.

There are two major groups of individuals, or more typically businesses or corporations, for whom the deployment of money must be seen as good value.

The first are the clients who commission the building. They have requirements that, starting with the function of the whole building, can be translated through the design to the function and then the performance of the parts so that technical choices can be made. If the building can be built for the money the client is prepared to allocate to the construction, the client is likely to judge the building to be good value.

The second group who have to have money as a resource to place at the disposal of construction are the professionals and the contractors who employ their money in the technical expertise and know-how of design and construction. Their attitude towards money, particularly the 'value' of long- and short-term investment, is crucial to the technical potential of the whole industry. If the human resource is just seen as employing people from the job market then this can be viewed as short-term investment. However, if, as is suggested here, the resource is not people but knowledge and skill then the investment is long term and its costs must be acknowledged. The willingness to deploy money in education and training fundamentally influences the technical capacity of the industry to deliver value for money to the clients as well as the long-term potential for their own money investment.

There is a third institution, government, that interacts with the investment and value-for-money decisions of the first two. Grants and tax regimes change the decisions of clients as well as the industrial players. There are also implications of the government's role in education, training and research and development to the investment attitudes of the industry. Not least is the role of the government as client, spending taxpayers' money on buildings. Not only can this be a sizeable part of the construction industry's workload but also, as the client, government can influence the way construction is organised.

Money is not a fundamental resource in the technical potential of a particular construction solution. This potential lies in the materials, the individuals with expertise and the machinery available. However, money plays a part in determining the levels at which these are available for construction and ultimately whether construction, however technically superior, will be applied.

Summary

(1) There are four resources that have to be available for the successful creation of a building: materials, knowledge and skill, machinery and money.

(2) Materials are the primary resource, being directly related to final construction of the building. Chosen for their properties to fulfil performance, they also have to be workable to make components suitable for incorporation into the building.

(3) The level of development in a society determines the ability to process and shape materials as it determines the growth of knowledge and skill, the application of machines and the availability of money.

(4) Knowledge and skill gained from both training and experience become invested in people. Analysis involves the quality of the understanding held by all the parties involved and the quality of the communication between them.

(5) Plant and machinery allow for the development of solutions that break out of the limits of individual human strength. The machines themselves have to be conceived and constructed and hence form a technological chain to provide the ability to use processed materials, manufactured into components, which can be erected on site.

(6) Money, although not used directly in the technical solution, enables the available resources to be brought together so long as society, the client and those who seek to earn their living in the construction of buildings see the endeavour as good value.

7 Design Concept

This chapter establishes the notion of an overall design concept for the building and its relationship to the function and performance of the elements of the building in order to inform the choice of construction.

Linking design to construction choice

It is not the intention of this chapter to presume to discuss either the theory or the practice of designing. This chapter seeks to make the link between the design and the choice of construction. The design will be an interpretation of the client's brief in defining the size and arrangement of spaces and their interrelationships, making them work in terms of their feel and environmental conditions for the operation of the building. The design will also have to respond to the physical and wider social context incorporating the requirements of regulation and the need to achieve sustainable development. This intention has to be worked right through into the specification and details to ensure the whole design concept is achieved.

When seeking a link to technological choice it is necessary to be concerned not only with spatial arrangements but also with flows and transfers in, out and around the spaces. Design is often seen as three integrated areas, architectural, structural and environmental, each with not only their own concern for space and aspects of the construction but also considerations for flow and transfers.

Designs therefore have to establish not only an arrangement of spaces, both internal and external, but also a way of defining this series of flows and transfers between them. It is by establishing the flows and transfers that the dynamic behaviour of the construction expected during the operation of the building can be visualised.

This allows an initial suggestion of the generic forms of construction that might be appropriate. However, perhaps most important, it allows for the identification of the function of the parts of the building and their contribution to the function of the building as a whole.

Flows and transfers can be classified into two groups. First are flows and transfers that take place through the construction itself. These are energy flows associated with environmental design, normally heat, light and sound, and transfers of forces through structural elements of the building. They also include physical transfer of gases and liquids through cracks and permeable materials.

The second group of flows is physical transfers through and around the building. Perhaps the most obvious from our experience of using buildings is that of people and objects that have to move through the spaces themselves. This has to be extended to considering emergencies such as fire, where an extraordinary direction and volume of people have to be defined. Further consideration of environmental design shows that energy often has to be transferred via physical active systems, in a medium such as air or water, or as gas or electricity that will require ducts, pipes and wires. In addition to these environmental systems, utility services such as water and waste disposal will be required for cleaning and operational needs. Other operational needs such as communications, alarms and transport systems such as lifts may be required. These active systems, collectively known as building services, are not just environmental. Similarly, not all environmental design is about building services.

The size and interrelationship of spaces, the fabric of the building and the building services and the interchange and transfers between them have to be expressed in a concept design. The design intention has to be made clear so it can be taken right throughout the process of choice of the technical solutions.

Concept design – general form

Architectural, structural and environmental considerations all need a conceptual design that can identify some generic construction forms or systems early in the design process which can be worked up into specific solutions later. Designs are made reality by technical solutions that can be conceived, along with the resources to put them into practice. Inevitably most designs are worked out with practical technical solutions in mind. This does not mean complete detailed solutions. It is more likely to be associated with some initial ideas as to the materials and sizes and even some details involved and how they would work with the other elements of the building. In some cases the demands of the brief from the client may lead to designs that cannot be achieved with immediately identifiable technology. This can be seen in the early sustainable designs using natural ventilation and with night purging schemes to reduce the energy consumption of buildings. This requires a deeper exploration of the general forms and the way they might work as the design concept is turned into practice.

All buildings pose challenges somewhere between copying the technology from the last building to developing new and novel solutions for general forms. Perhaps the key understanding in this process is just how difficult it is to achieve a technical solution to an evolving design. There is no direct relationship between novel designs and technological difficulty. The radical design may be easily achieved with existing technical knowledge and experience.

Physical performance and appearance – commodity, firmness and delight

The design concepts must allow for the function of each part of the building to be identified with the level of performance required if the building is to function as a whole. The move from client's brief with its language and images of the operational nature of the building to a definition of the technical requirements of the parts should emerge from the design concept expressing the intention for both physical performance and appearance.

Concepts of appearance are a matter of taste, style and image. They involve aesthetic judgements and rely on responses conditioned by cultural and social influences. Appearance should not be interpreted too narrowly for this aspect of design. It involves not only the finish of the surfaces that can be seen but also the sense of space and the appearance of the space itself to the beholder. It is as much (if not more) to do with the feel of the space defined by the building as by the strict look of the building itself.

The business of providing a working building is a matter of physical performance. Concepts have to be based predominantly on the realities of the physical world and the laws of nature, although the arrangement of physical space influences our psychological well-being. They involve physical, environmental and structural judgements and have in the recent past called on knowledge from ergonomics, science and engineering. They require an understanding of the physical behaviour of the construction when subjected to the forces of nature.

Most definitions of 'good design' would argue that both the working performance and how the building appears should be satisfied in order to attribute any judgement of quality to a building. The Roman architect Vitruvius expressed this in the first century as commodity, firmness and delight. Commodity and firmness refer to the building's ability to accommodate the user's activities and to its structural integrity. These have been referred to earlier in this text in terms of utility and safety. Both of these require attention to the building's physical performance. Delight, on the other hand, is about the way individuals feel about the building. Without all of these, the building is left wanting in some respects. All are legitimate objectives to be set for the technical solutions to achieve. However, since the nature of these objectives and subsequent evaluation have to be made in different ways the distinction between the physical performance and what is being referred to in this text as appearance (delight) seems appropriate here.

Appearance and physical performance differ in the way judgements of success are made. The response to aspects of the appearance will be based on a human, cultural, even spiritual response expressed as emotions, if they can be expressed at all. The response to the physical performance will be physiological and psychological, expressed in feelings of comfort, and a mental state. There will almost certainly be an overlap between the two. Both the appearance and the physical performance in aspects such as safety and privacy influence a sense of psychological well-being.

Function of the whole, function of the parts

The design is the translation of the client's physical and social needs for the operation of the whole building into a scheme that must be able to identify the function and performance of each of the parts. The conceptual themes that unify the appearance and working of the building have to be able to identify the role of each part of the building in achieving the performance of the whole.

The design will have to identify not only the function but also the performance level. For example, the environmental design will determine whether a wall will have a function to resist the passage of heat. The design will have to determine the required internal condition and then will have to suggest the relationship of heat inputs to retention rates and the role of services and passive construction in maintaining the design conditions. Given the general concerns for energy saving, most designs look for retention of heat to balance low energy inputs which means a high performance requirement from the wall in resisting the passage of heat.

Having established the basis of the function and performance levels of the elements, assemblies and components of the building it is now possible to undertake the analysis necessary to evaluate proposed construction specifications and details.

Summary

(1) There has to be a way to link the design concept formed from the client's brief to the process of technical choice of the elements, assemblies and components of the building to ensure that the design concept and intention is carried right through to the specification and detail.
(2) The design has to determine not only the arrangements and interrelationship of the spaces themselves but also the flows and transfers required between the spaces.
(3) The design is most likely initially to be developed with some ideas about the generic solutions, from which the specification and detail could be developed to ensure the full design intentions are achieved.
(4) Good designs require both physical performance and qualities of appearance (delight) which have to be addressed in the technical analysis but have to be considered separately.

8 Appearance

This chapter suggests the ways in which appearance may be interpreted from the design concepts and then introduces how these influence the physical construction and the space that may be available for the technical solution.

Visualising attributes and response

It was suggested in Chapter 7 that appearance is perhaps an inadequate word to convey the full meaning of this requirement of the design concept that has to be achieved through the specification and detailing of the construction. The words 'feel' and 'delight' were also used, but even this needs to be explored further if it is to be successfully translated into the choice of construction.

It is necessary to think about the attributes that will influence the appearance of both the space and actual physical presence of construction. There is a need to judge the response of people to a suggested solution. The response will be conditioned by cultural and socialising influences. The response required may be an aesthetic judgement concerning taste and beauty. It may also be required to convey some message about the purpose of the building or the status or beliefs of the owner. These would have been part of the brief and should be present in the design concept. It is then important to evaluate the extent to which the suggested solution evokes these appropriate responses.

Space and context

The requirements of appearance will be primarily judged by our response to the experience of the building and the spaces it creates. Buildings divide space: internal and external, public and private, light and shade. These not only must inform and enhance the function of the building itself but also should respond to the context of site and surroundings. These are deep responses and involve judgements with cultural and social overtones. This generates a paradox in that while they may be the most difficult to articulate objectively, at the time of choosing the solution it is the attribute that is most easily 'measured' by the users of the building.

Since this book is predominantly technical it is not intended to discuss this analysis in any detail. It may, however, be of value to examine one or two of the issues that contribute to the formulation of this design intent from the brief and the context in which the building is to be placed. The client's brief has to be translated into appropriate spaces arranged to mimic the activities in both scale and interdependence. In achieving this the design intent has to encompass such ideas as style, image and aesthetics as well as context.

The need to satisfy the requirements of appearance also influences the space available to the construction since it can only occupy the zones between internal space and external form. The design concept can be confounded not because a construction form cannot be conceived and created to perform but because there is not sufficient space for it to exist without it adversely influencing the aesthetics. All these are the ideas that are briefly explored in the remainder of this chapter.

Style

Historically many cultures have developed different styles of buildings. To the Greeks the analysis was predominantly geometrical. Early in the twentieth century modernists saw buildings as representing order and clean living. Later in the century many buildings can be seen as expressing the overt use of new technologies. Everyday building, where the form has evolved driven by local conditions, is often referred to as vernacular.

What is perhaps interesting in this historical perspective is that, however different the cultural origins, there is considerable agreement that, in their own way, all of these buildings can be of great quality when the style is carried through to the specification and detailing. It is perhaps the unity of concept with careful attention to ensuring it is maintained right through to the details in the construction that is important.

Copying style is often not successful. There does seem to be a need to thoroughly understand the origins of a style, to be driven by the same unwritten forces, and to appreciate the objectives it satisfied for the owner of the original buildings. It also seems important to have the appropriate resource base. The very quality of the materials and the understanding and skill of the craftsmen which developed in the exploration and development of the original style will be missing in any attempt to reproduce or copy a style in another place at another time.

Image

Another aspect of any design concept is the image it intends to portray. This is associated, not directly with the activity, but more with the social and cultural purpose of the building and the dreams and aspirations of the owner or occupier. It involves issues of status and statements about social and commercial position. It puts forward the social (or individual) beliefs to which the owner/occupier pays particular regard.

This holds true if the owner is not an individual but corporate or governmental, although often in this respect the 'image' is the creation of one individual, most often the head of the corporation or government. Commercial, business and political beliefs dictate the appearance of buildings. Compare the civic pride of the Victorian town halls with the bureaucratisation of local government manifest in the public building of the late twentieth century; or the sense of security in banking halls when the business was holding clients' money with encouragement to deposit, compared to the open layout when the banking business became predominantly credit and lending. Both these examples demonstrate a considerable shift in the importance of space and the choice of materials and detailing.

Although the purpose of two buildings may be the same and hence the functions of the building remain constant, issues of image may determine the spatial arrangement and materials quality that significantly influence the technical solutions. It may call for larger or smaller spaces which will influence the possible choices of structural solutions. Image determines aspects that will influence the choice of quality materials and have considerations for life spans of components. Materials may be chosen not only for properties of appearance but also for their ability to be worked to fine limits or their qualities when weathered.

Image can become stylised into building as the major vehicle for corporate image. Once this state has been reached many design decisions are removed from the process. While spatial questions and best use of site will still have to be considered, the building may become a series of standard details, materials and colours with only limited scope to influence the design. If this is carried through to solutions based on prefabrication and off-site processes then decisions on design are even more limited. This can be seen in franchised retail outlets, particularly those associated with fast food at the end of the twentieth century.

Often the ideas of image are not directly translated into the spatial arrangement and the appearance of the building. The image of the 'green' client, 'caring for the planet', may introduce, through the brief, consideration for choice of materials and details in providing the physical construction of the building. This fact may or may not then be used overtly by the client at a later date depending on whether the image is significant to the client's business or whether it is just part of the client's self 'image' as part of his/her own personal beliefs and aspirations.

Aesthetics

Beyond the issues of style and image, yet often incorporated within them, lie the pure ideals of aesthetics. Concerned with the recognition of beauty, it is a quality sought in much art, of which architecture is a recognised form. It is concerned with feelings generated within the beholder when confronted with beauty. It may not be too strong to suggest that the experience can be spiritual. The human response in this case may defy analysis, but the emotions it creates, and the influence it has on well-being, would be shared by all of similar taste that experience the building.

Response to context

Buildings are experienced not only from inside, associated with the activity of the building itself, but also by a wider public from the outside. In addition, the view from inside a building can be an extension of the internal space itself and change the experience of the internal space. Buildings contribute to landscape and cityscapes, in which they may enhance or detract from the sense and feel of the space.

Some aspects such as orientation to provide logical access and movement between inside and outside or to provide the best use of the natural environmental conditions such as sunlight are relatively easy to identify, analyse and justify. Other aspects such as scale, massing and the sympathetic use of materials are often not so easy to identify, particularly in the context of a statutory planning framework.

Buildings can become the subject of public debate locally and in some cases even nationally. Such debate often centres on controversial responses of liking and disliking, sometimes the building itself, but often its relationship to its surroundings. These debates are normally about the shock of the new or the tedium of the widespread. It is often not clear whether the building itself lacks design content or if it is its position and surroundings that are in contention. What does seem to be clear is that both need to be agreed before the building is to attract universal acclaim.

Physical space for the construction

The bulk or slenderness of the construction has a profound effect on its appearance. The size and scale of the actual fabric of the building and the detailing will have to be in sympathy with the intended style, image and aesthetics as well as being significant to context. This is true for both the internal spaces and the external appearance of the building and its contribution to its surroundings.

Much of the expression of appearance is about aspects such as line and surface shape as well as colour, texture and light of the actual surface. These are spatial limits to construction as well as definitions of the surfaces themselves. The construction has to be contained within a zone that remains when internal space and the external form have been defined.

This limit to the size and arrangement of the construction has to be part of the analysis. It may be that in order to achieve the required level of physical performance in an economic manner, the size and shape of components will exceed these limits and intrude into the visual field of the onlooker, interfering with the required appearance.

Moreover, these spaces or zones for the physical construction often have to contain a number of elements of the fabric and services required for the building. Planning for the space for the elements of construction can become as significant as planning for the space for the occupants.

The approach to analysis

Image, aesthetics and response to context are not subject to the same type of analysis as the physical performance expected of the construction solutions. They do, however, influence the choice and must be part of any evaluation of a suggested solution. The analysis is not a dialogue with the laws of nature but a dialogue with the perceptions of taste and cultural norms. When these are not well defined, design is often seen to lose its way and levels of controversy over buildings grow.

Normally, much of the construction is hidden from view and therefore space availability is the only limit associated with appearance. Clearly some aspects of construction, referred to as the finishes, are specifically required to contribute to appearance. In some designs the natural aesthetics of form of the components is used as part of the design concept. In these cases, shape and size and in particular joints and fixings become part of the generation of a desired appearance.

Some designs employ decoration and appendage to create the desired qualities of appearance. These parts of the construction may play no other part in fulfilling function other than appearance. Their justification lies entirely in the contribution they make to the response they invoke, giving value to a particular form of style and aesthetic expression. No less than any other part of construction, choices have to be made about the materials, shape and size plus some means of fixing. This is still a technical challenge for the parts still have to have sufficient strength and need to maintain their effect over the years. In some forms of decoration such as sculpture, the artist determines material size and shape; only fixing may become part of the choice for the construction.

The ability to generate a building that fulfils these criteria would be associated with a creative act in the best traditions of art. Historically this may have been more associated with craft where the emergence of design came from experience with the materials, rather than an analysis of the creative arts themselves. This represents a technological tradition where design, detailing and construction were more closely related in the everyday choice of technical solutions for buildings.

Summary

(1) Appearance is a major aspect of the design, associated with evoking response to the building conditioned by social and cultural norms.
(2) It concerns the generation of space as well as the physical surfaces and ornamentation and is associated with the external appearance in response to context as well as internal design.
(3) Appearance can be associated with the creation of style and image as well as aesthetic consideration.
(4) This makes demands not only on the choice of materials and details but also on the space that is available for the construction between the internal design and the external appearance.

9 Analysis of Physical Behaviour

This chapter is the introduction for the three chapters that follow. It establishes the rationale for the titles of these three chapters and draws out the common approach to analysis that is associated with understanding physical behaviour. It considers function and performance as the outcome of the physical behaviour, which, linked to the need to prevent failure, leads to the need for an explanation of how the construction responds when working under design conditions.

Visualising conditions and response

Given that a proposal for the construction has been suggested, however tentative or outlined the description, it is possible to start to think about its physical behaviour. There are three primary areas of behaviour that are of significance to the success of the building, namely:

- Behaviour creating environments
- Behaviour under load
- Behaviour over time

The analysis of physical behaviour is predominantly a technical activity since it is undertaken after the function of the parts has been defined and after the conditions under which it has to operate have been established. It varies from an analysis of the appearance of the building in that it is not based on a social or cultural response but on a response to the condition of the physical environment. The analysis of physical behaviour must be based on an understanding of the laws of nature; the forces involved both in states of equilibrium and in the dynamic behaviour during changes in conditions.

Function and performance – a technical aim

This analysis of physical behaviour takes as its criterion for success the idea of performance. Every part of the construction fulfils a function. It has a reason for its existence and a part to play in contributing to the function of the building as a whole. While the client or user determines the function of the whole building, the functions of the parts are determined by the chosen design concept.

Having established a design and assigned functions to a part of the construction, it is necessary to identify a level of performance. It is against these levels of performance that analysis can be undertaken and the failure modes identified. The language of performance levels will vary, but the common idea is that the performance statement must be testable. Examples would be that beams must carry load without exceeding a determined deflection, walls may have to resist the passage of heat with a specified level of insulation, and no dampness should appear on any internal surfaces of the external elements of the building.

Preventing failure – a technical objective

Failure is a loss of performance. Failure follows when the response of the proposed construction does not meet performance levels. If the beam is too small or made from a material that is too elastic and has not been shaped or fixed to overcome these shortcomings, it may deflect too much. The beam has failed to perform. If the loads were larger than predicted and the beam deflects too much, the origins of this failure are in the prediction of the design condition.

The analysis of physical behaviour requires three areas of understanding:

(1) Identification of all the performance criteria
(2) Careful consideration of the source and nature of the operating conditions
(3) Understanding of the mode of failure of the construction responding to working conditions

This requires an ability to imagine the ways in which the construction might physically fail to achieve performance under the design conditions. The clues to this should be in the performance specification itself, for example, excessive deflection in beams under loading conditions or excessive heat transfer through the walls under winter environmental conditions. It will also require a method of predicting how much deflection or heat transfer will occur to check that behaviour will remain within performance limits. The clues to this will come from visualising the internal action of the construction under these conditions in order to determine the response.

Future predictions play a part in this analysis and this is often difficult, given the expectation of longevity for most buildings. Just establishing higher performance levels for some expected future requirements is problematic, for higher performance levels will have cost implications. Variations in conditions with time, including accidental events, need to be predicted with some probability, since designing for all imaginable circumstances has considerable cost implications.

These considerations for the future add to the difficulties of prediction. They may be included in the brief but, if not, can be part of a technical analysis in so far as the sensitivity to error or change in conditions can be carried out. A solution that is sensitive to change may be less attractive if the future is not certain. This adds an additional dimension to the analysis. How close to failure is the solution when under working conditions? Will small changes greatly increase the probability of failure?

This prediction of sensitivity can be part of the analysis undertaken by visualising the response of the construction under the expected conditions. Understanding of the response of the construction to the external conditions allows the estimate of behaviour under working conditions and any capacity it may have to withstand unexpected or deviant conditions.

The role of observation and science in the analysis

It is suggested that an understanding of the ways of nature is necessary to provide the descriptions and determine behaviour and hence the performance of the construction.

Experience can offer direct observation of construction both during production and under working conditions. In unfortunate cases experience can offer the opportunity to observe failure. If no failure is observed in a form of construction it can be repeated in similar conditions and be expected to achieve similar performance levels, although this observation tells us nothing about the sensitivity of the construction. If failure is observed then some explanations are necessary to determine what changes are required to achieve performance the next time. Requirements for higher performance, operating in different conditions, or a change in the construction for production for economic gain also require explanations. These explanations have to be in the form of a description of the response of the construction in resisting failure.

For those engaged in technological activity the natural enquiring mind is likely to seek such explanations. It is possible that over the past centuries, individuals have provided their own explanations in order to develop construction in a systematic way, refining solutions and occasionally making larger leaps of understanding to new forms of construction. Whatever the scale of change, these explanations have given confidence in perusing new ideas that would achieve performance with a limited (or at least considered) chance of failure. It is interesting to speculate whether the great cathedral builders had their own explanations to visualise behaviour.

Science provides a framework for exploring the laws and forces of nature. It provides a method of investigation from which explanations of the natural world can be verified. These explanations are useful in explaining the behaviour of construction. Science does, however, have its own language which can be daunting to many. The knowledge derived from science has two major powers, one of explanation and one of prediction. Only the second, that of prediction, really needs to use the language of science. Explanations can be reduced to everyday language for they are only descriptions of the way the laws of nature operate. While it is the explanations that help understanding of the way in which construction may be functioning (action), prediction becomes necessary to establish a level of performance.

The approach to analysis

The analysis will be based on visualising the dynamic actions and processes involved. This will be achieved predominantly with verbal explanation or word pictures. Description of the mechanisms involved seems to offer the most widely applicable approach to understanding technology. All this can be greatly assisted by diagrams and mathematical modelling on the one hand and by experience on the other, but both these seem to need a verbal articulation,

not only for a personal understanding of behaviour but also to provide an explanation to others of the basis on which the solutions have been adopted.

The aim should be to build a picture in the mind's eye first of what the solution will physically look like (the suggestion) and second of the actions that determine the response that describes the potential failure under working conditions (the analysis). Providing a description of both physical construction and then the response of the construction against criteria that, if not satisfied, would be considered failure is the underlying approach used in this text.

Four basic areas have been established and need to be understood to complete an analysis:

- Conditions expected through the life of building
- Failure or loss performance
- Proposal or suggestion of the construction (components and assembly)
- Response of the proposal to the condition

The chapters that follow provide the basis for this visualisation. It involves a dialogue between the proposed construction and the response of the construction to the conditions in which it has to be built and operated in order to limit the risk of failure.

Summary

(1) Performance identifies the level to which the construction has to function under the design conditions and becomes the criteria for any evaluation of success.
(2) Failure is a loss of performance at some time in the service life of the building under a predicted set of conditions.
(3) Failure follows the action of construction. Action is the internal response of the chosen construction under the conditions in which it has to perform.
(4) Visualising the conditions under which the construction has to function and the expected response of the construction chosen provides the basis for an analysis to evaluate satisfactory performance.

10 Physical Behaviour Creating Environments

This chapter takes in turn each of the aspects of environment that are important to the well-being of individuals and their possessions in the building. These have been identified as dryness, warmth, light, acoustics, cleanliness, security and privacy. In achieving these environments the building must be seen as a system employing both active services and the passive building fabric. The overall performance requirements will depend on the severity of the external conditions and the specification for the internal working conditions. The design will have to assign the role and contribution (function and performance) of each part of the construction to the creation and maintenance of the required environment. Once the contribution to performance for any part of the construction is known then it is possible to undertake an analysis of the response of construction under the working conditions to evaluate the chance of failure. For each of the aspects of environment, the dynamic response of the proposed construction can be visualised based on an understanding of the law of nature and hence the action or mechanisms that might lead to failure.

Aspects of environment

This chapter is concerned with the aspects of physical behaviour of the construction that ensure the internal environment required to support the activity carried out in the building. These activities require aspects of both social and physical environments to be considered to ensure the well-being of the occupants and their possessions. All aspects of the physical environment will rely on the physical behaviour of the construction to ensure a satisfactory internal environment. However, not all aspects of the social environment will require a contribution from the physical behaviour of the construction. Space and appearance is not considered in this text as a physical behaviour and has been discussed in Chapter 8. Some aspects of the social environment such as privacy and safety do rely on some physical response from the construction. In the case of safety, the physical condition may be associated with loading and will, therefore, require the analysis outlined in Chapter 11. The range of physical and social environments is given in Figure 10.1.

The building as a system

Before evaluating the behaviour of the construction it is necessary to assign functions to each part and define the required level of performance; that is, the contribution that each part has to make in ensuring the conditions required by the user in the building as a whole.

A definition of the required environment does not say anything about the contribution to be made by the various parts of the building. Even when the social and physical worlds outside the building have been defined, all that has been established is the total task of modification.

Before functions can be assigned to the various parts of the building it is necessary to develop an overall view of how the required environment is to be created and maintained. There are three basic groups of technologies available to fulfil these functions:

- Passive fabric
- Active fabric
- Active services

These are shown in Figure 10.2. It may be argued that the supply and disposal services are not

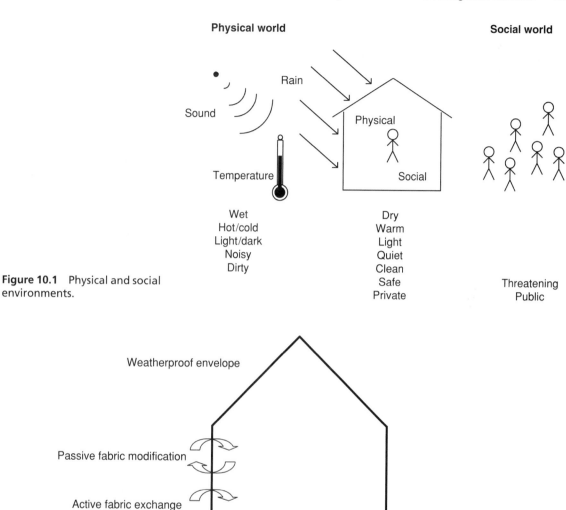

Figure 10.1 Physical and social environments.

Figure 10.2 Creating internal conditions – possible technologies.

all strictly environmental services but all contribute to the provision of the active creation of the environment even though they contribute to other operational services as well. Water, gas and electricity, for instance, are often categorised as utility services.

Many of the aspects of environment can be achieved with different combinations of these three types of technology; for example, the use of artificial lighting to supplement natural lighting provided by windows or roof lights together with the colour of walls. Decisions on this balance are influenced by initial costs, running costs, energy use and the control arrangement to be incorporated. With an increase in emphasis on the environmental impact of buildings, the combination of building design and active fabric solutions can reduce energy use from the active services.

Even when this balance between services and fabric has been established it is still possible to assign the functions of the fabric in a number of different ways. Some functions are fairly obvious, such as the roof covering's function of weatherproofing to create and maintain a dry internal environment. For others, alternatives exist. In providing a warm environment thermal insulation may be chosen to maintain temperatures in order to limit the need for heat inputs. This can become a function of either one or all of the main elements of the building, roof, walls and ground floor, each with various levels of performance contributing to that part of the strategy for maintaining internal temperature levels.

When the balance between services and fabric has been established, it is possible to start to develop details and specifications. The approach to each of the three types of technologies is different, generating a different set of decisions for each.

Passive and active fabric

Where the creation of the internal environment depends on the contribution of the fabric, the process of choice is focused directly on the variables identified in Chapter 4.

If passive fabric is to be employed then the primary concern will be for:

- The properties of the materials
- The dimensions, usually thickness
- The joints and fixings

For example, if sound transmission is being considered and mass is to be used to achieve performance then it will be seen later in this chapter that the density of the material will determine the thickness required, so long as the joints do not support flanking paths and the fixings provide sufficient rigidity.

Where active fabric is being used then the focus becomes the joints and fixings to ensure the operation of the moving parts. The material, shape and dimensions of the active component will be determined more by their strength and reliability in use, although they may also have a passive contribution to make that will require analysis, as suggested above. They may have to be considered for appearance, as these components are almost certainly visible.

For active fabric there will also have to be some form of control so that their effect can be brought into use when required.

Active services

The final choice of components and materials in services systems is not as straightforward as the direct choice that associates the specification of the fabric with the environmental conditions. Their choice has become a design discipline in its own right. They are systems within the system of the building as a whole. Figure 10.3 suggests a way of considering the active environmental systems and how they might be integrated into the overall building design.

The active services are required to link the user with either a supply or a disposal system, as shown at the top and bottom of the diagram. Taking a heating system as an example, a radiator may be provided as an outlet but has to be linked to a heat source. In this case the active distribution would be via heated water contained in pipes, most probably using a pump to provide a mechanical force. The radiators would be located in each room probably under the windows. The boiler to provide the heat would need to be located to make fuel and flue connections plus with consideration of appearance and as a possible noise source. The pipes may be routed through the floors and either within or on the walls of the building or through specifically designated ducts or voids. This would have to include an access point to allow for repair and maintenance. In another example, a natural

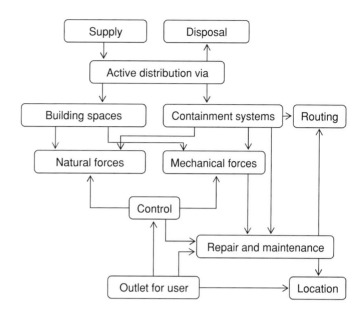

Figure 10.3 Aspects of active services systems.

ventilation system might use the building space as the distribution system, using natural forces of wind and thermal gradients to move the air from the user, to dispose stale air at high level and bring in fresh air (perhaps tempered with heating) at low level.

Figure 10.3 also shows that, like the active fabric, these active services systems require control. When the internal conditions change, some control mechanism will have to be invoked. This control can be manual, relying on the user to identify unsatisfactory conditions and then being offered some way of re-establishing the required environment. For example, opening a window as long as fresh and relatively cool air existed outside the building could control a stuffy or over-warm room. It would include leaving the occupants to switch on lights when they considered it to be too dark. The alternative is to devise ways of providing automatic control, such as the use of a sensor to respond when design limits are exceeded and send a signal to a part of the system, which can start a chain of events to bring the environment back into design conditions. In the example of the heating system in the previous paragraph, this might be a room thermostat that can switch on the boiler and the pump in a central heating system when the

temperature in the room (more correctly at the thermostat) falls below a predetermined limit.

Assigning environmental modification functions

Table 10.1 shows the aspects of the internal environment that may have to be created and maintained. These are each then identified with the major aspects of the external environment that will influence the achievement of modified conditions. The final two columns indicate typical functions that might be assigned to parts of the construction to achieve the modification required. For most of the aspects of modification, both passive construction and active services could be employed with different functions. For example, to create dry internal conditions it is necessary to understand patterns of precipitation, the conditions of ground water and the presence of vapour in the air. From this, waterproofing can be assigned to the fabric and water disposal to the building services.

Rarely does one part of the construction act alone in fulfilling one of the environmental functions and rarely does it contribute to one of the functions only. Elements of the building such as walls and roofs may contribute to many of

Table 10.1 The role of fabric and services.

Internal conditions to be established	Significant aspects of the external environment	Possible functions of fabric	Possible function of services
Dry	Precipitation Ground water Vapour in air	Waterproofing Vapour control	Water disposal
Warm	Air temperature Solar gain Wind Humidity	Resisting the passage of heat Thermal mass Ventilation	Heating Cooling
Light	Sun path Overcast sky	Transmission Reflection	Room Task Mood Architectural
Acoustic	Natural sounds Pollution	Transmission Absorption	
Clean	Dirt Pathogens Fungal and plant life Animal life Air pollution	Exclusion Filtering Cleanable	Removal Filtering Disposal
Safe	Social threats Act of God Fire	Resistance Separation (barriers)	Detection Alarm
Private	Cultural and social conventions	Separation (visual and acoustic)	

the functions. Elements are assemblies of sub-elements, such as the external skin of a cavity wall or tiling on the roof, where a primary function is often identifiable. Choice often centres on these sub-elements, for they are normally identified with a major component or material where the focus of the functions is also often fairly clear. For example, an external wall probably plays its part in all the environmental functions. Even major sub-elements such as windows have a range of functions. Primarily for light and view, these may be used successfully for ventilation but cause problems with solar gain and security. As finer division of the sub-elements is considered, down to single components, then function is often clear. Within a wall, the damp-proof course (DPC) has a very well-defined function of damp control. Analysis at this level has to keep sight of the function of the building as a whole, while defining the function of the components and materials of which the sub-element will be made. This division and refinement of the

environmental functions is illustrated for a roof in Figure 10.4.

Assigning performance levels for environmental modification

Assigning of a function only identifies the part to be played by an element, sub-element, component or material in creating and maintaining the required environment. Assigning a performance involves defining a level of attainment. While it is unusual for a building to be of such a radical design to cause the functions to be unclear, the new demands of clients and regulations often lead to a change in performance. All the aspects required to set a performance level have already been introduced:

- The range of what will be considered as normal external conditions and what extreme or accidental conditions it may be worth designing for

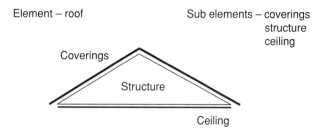

Figure 10.4   The role of sub-elements of the roof.

Contribution of sub-elements to environmental performance

Coverings	Dry	Waterproofing
	Clean	Infestation exclusion
	Safe	Fire spread
Structure	None	
Ceiling	Warm	Resistance to passage of heat
	Light	Reflection
	Clean	Dirt exclusion

- The ideal internal conditions, the limits that would still be considered satisfactory and any control mechanisms that are intended to be included
- The proportion of the task of creation and maintenance of the environment to be assigned to that part of the construction

Thinking through the conditions

The analysis of each aspect of the internal environment will be considered in turn in the rest of this chapter. The focus will be on the analysis of the fabric, although the choice of services systems will be introduced.

Following the general format for thinking through physical behaviour established in Chapter 9, each section will consider:

- The origins and nature of the environmental conditions
- The nature of the failure linked to performance
- Possible types of solutions
- Analysis of the response involved to achieve performance

A dynamic description of the environmental conditions to be modified and the mechanisms that determine the response of the construction to the expected environments in which it will have to operate should be carried out. This should form the basis for a visualisation of the action of any part of the construction deemed to have a performance requirement and how it might fail in contributing to ensuring the internal environments.

The dry environment

Substances exist in a solid, liquid or gaseous state. Unlike nearly all other substances, water can exist in all these states within the normal climatic range of temperatures and pressures. This creates particular problems for buildings trying to maintain dry environments. An excess of liquid water is undesirable, but a limited quantity of gaseous water (vapour) in the air is important for comfort, providing levels of humidity. In visualising the behaviour of construction, the possibility of a change of state has to be considered. Perhaps the most important example of this for dry environments is the change from vapour to liquid in the depositing of condensation.

As well as having an undesirable effect on health and comfort, water is a major source of deterioration of construction materials (see Chapter 12). Much detailing has the dual function of protecting the environment and the fabric of the building. In this respect, visualising the dynamic interaction between water and the construction is common to the analysis of both environmental control and durability. However, the analysis of the two need to be considered separately since different actions of the water become significant, as, for instance, in the change of state from liquid to solid with frost action, significant for durability but of little concern in considering environmental conditions.

If the requirement is for a dry interior then the clear failure can be characterised as leaks, where water enters the building through the external fabric and openings or from failure of the building services. However, much of the detailing of buildings is concerned not only with the free entry of water as leaks but also with excessive moisture in the fabric, known collectively as damp. The dividing line between these, both as observable actions and in the appropriate detailing, is often fine, so they are normally considered together.

Maintaining a dry interior (absence of dampness rather than lack of humidity) requires the control of moisture as both liquid and vapour. As a liquid, the source of water is normally external and hence the external fabric has a major functional role in damp control. As a vapour, sources are predominantly internal so both the design of the external fabric and the overall process of ventilation are used to prevent the incidence of condensation.

When considering the analysis and choice of the fabric to ensure a dry interior it is usual to see three major threats:

- Penetrating damp
- Rising damp
- Condensation

Among others are blown light snow entering ventilated openings and subsequently thawing and leaking from water-carrying services.

To visualise the behaviour and analyse the ability of the construction to fulfil the function of damp control requires knowledge of the origins of the moisture, the forces acting on it when it comes into contact with the construction and the pathways then available for it to reach internal surfaces.

Predictive methods are available to model vapour and temperature patterns in order to assess the risk of condensation. Penetrating and rising damp assessment is still largely a matter of traditional detailing; maintaining barriers to the progress of dampness into a building by applying tried and tested methods. Testing is possible of prototypes to wind and rain penetration, particularly associated with components systems such as windows. This prototype testing is especially significant when new principles such as rain screening are applied because predicting the behaviour of water can still be a matter of trial and observation.

Each of the major threats will be considered in turn, with descriptions of behaviour, identification of pathways and of the major detailing techniques that can be employed.

Penetrating damp

Rain is the major source of water that interacts directly with the external fabric of the building, creating the possibility of penetrating damp. Three major actions can then be identified, as follows:

(1) The water that comes into direct contact with the outside of the building will soak into permeable construction where pore networks form continuous capillary pathways through the material.

(2) Any water running over the surface of the building is primarily influenced by gravity moving it down the construction, but other forces also act on this surface water, particularly when it meets joints and projections. The water may be drawn into small cracks by capillary action. Wind that accompanies the rain will create pressures around the building, affecting rates of movement of the water across the surface and penetration of the fabric particularly at open joints. Even the 'skin' formed on the surface of water exerts a

force known as surface tension that not only holds water on the surface but also can cause it to migrate, upside-down, under horizontal surfaces beneath overhangs. Visualising this highly complex set of forces acting on water on the wetted surfaces of buildings is particularly important at the detailing of joints between components and at junctions between elements.

(3) The majority of the water that falls on the surface of the construction will arrive at an edge. If it is a free edge, the water will fall off mainly under the influence of gravity. However, its behaviour at these edges will also be influenced by wind and surface tension drawing some of the water back under horizontal overhangs. It is important to make sure this water is removed from edges effectively so it does not have the chance to accumulate and soak the construction more than that expected from the original rainfall.

In reality the distinction between the first action (soaking in) and the second (surface water interacting at joints) is not so clearly defined. Most construction, such as roof tiling and brickwork, is not monolithic. It is not just one material where the properties of the material will be the only influence on the rate of soaking in. Although they may be considered as one skin they are made up with joints which in the case of brickwork are made with a second material, mortar. The behaviour of 'soaking in' is now dependent not only on the permeability of the brick and the

mortar but also on the reaction of the water at the joint. Some of the forces identified in second action (interaction at joints) come into play. In fact, in brickwork with cement mortar, where drying shrinkage breaks the bond between brick and mortar to create fine capillary paths, water penetration can be observed at the joints long before it penetrates the brick itself.

The depth of penetration of the water soaking into the material depends primarily on two things: the cycle of wetting and drying and the permeability and thickness of the construction. Prolonged periods of wetting with only short periods of drying in between pose the greatest threat. If the construction is not specified thick enough, or barriers to halt penetration are not provided, moisture will appear on internal surfaces. The three basic construction forms to overcome this direct transfer through the walling material itself are shown in Figure 10.5. All of these forms can be exploited. At the beginning of the twentieth century most house construction would have had solid walls, but by the end of the century, the changes in materials, particularly the use of cement mortars (see explanation in previous paragraph), had led to the almost universal adoption of cavity construction.

The water that does not soak in or is not drawn in will move down the face of the building. If the surface is impermeable then a considerable quantity of water can accumulate. There is a danger of damp penetrating at any time when the water meets joints or projections. The expected pathway and potential quantity

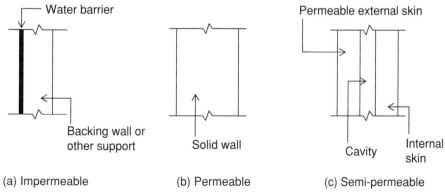

(a) Impermeable   (b) Permeable   (c) Semi-permeable

Figure 10.5   Generic forms of walls to resist penetrating damp.

Table 10.2   Forces potentially assisting water to enter at joints.

Force	Direction of action	Effect
Capillarity	Any	Water travels through fine cracks and pore networks
Gravity	Downwards	Water enters any opening or crack with downward path
Pressure differences	Any	Wind pressure
		• Driving rain
		• Holding water in joint
		• Moving towards lower pressure
		• Pumping behind flashings
		Water pressure below water table in ground
Kinetic	Rebound	Localised to area of impact
Surface tension	Any surface	Movement at underside of projections
		Holds film of water on surface and on ledges

of water has to be visualised for each situation of joint and projection.

The action of capillarity at a filled joint has already been identified between the brick and its mortar if the bond breaks and fine hairline cracks are formed. This is the scale at which the mechanisms have to be visualised. Pathways through the fabric are often very fine and tortuous, with damp appearing on the inside, some distance from its entry on the outside. Other joints are open with no filling material. If these joints are narrow they may support capillarity if a fine continuous open gap is formed. However, two other forces will assist water penetration at gaps or joints: gravity and wind. These may be significant even where the gap size would not support capillarity. Not only may wind change the surface water's movement directly, but also the pressures that it builds up around the building may blow water through joints or even pump water up behind detailing such as flashings.

Two other forces that may act on the water when at joints or projections are kinetic and surface tension. Kinetic energy is present in the falling rain. It is the force that causes it to bounce on hitting the surface, so apparently sheltered areas may become wet. Kinetic effects are relatively weak, but surface tension has the potential to cause problems at overhangs and projections. On the underside of projections, which may also be considered to be sheltered from wetting, surface tension will hold water on the underside of the projection where it will migrate towards

an apparently protected joint and enter from below the projection. The basic forces acting at joints are shown in Table 10.2.

Penetration at joints can be aggravated when they occur at projections such as sills. Water will collect on the top surfaces of a projection, causing high levels of soaking or additional water to be drawn in through joints. Identification of the damp pathways and the detailing of damp-proofing at the sill of a window is shown in Figure 10.6. This figure illustrates the use of jointing materials (putty), filled joints (water bar), barriers (DPC) and, on the underside of the sill, the simple device of the groove or drip detail. In the drip the water collects and gains sufficient mass to break the surface tension when droplets fall from the underside before reaching the joint under the sill.

The distinction between absorption from prolonged wetting and direct penetration through joints and projections has particular importance when thinking about aspects of external climate and rainfall in a particular locality. Any building subject to heavy, driving rain may need to be detailed differently to one subjected to prolonged drizzle. The total average annual rainfall may be the same.

Finally, it is necessary to consider the water that is shed from the building particularly where the roof discharges water at the top of walls. If not intercepted, gravity will take it towards the ground. It is not a direct threat if it falls directly to the ground, but it may fall or be blown

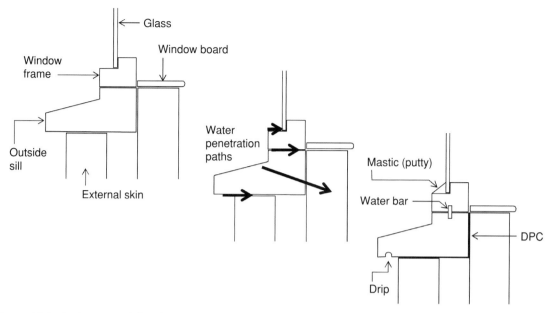

Figure 10.6   Pathways and details to resist damp penetration at sills.

onto the surface of the construction, soaking materials to an extent not envisaged by direct exposure to the rain itself. If it reaches the ground without contact with the construction it may still splash (kinetic forces) onto the fabric at ground level and increase wetting and soaking of materials close to but above the ground. The rest of the water reaching the ground will either directly soak into the ground or collect and flow away on any flat, hard surfaces. Intercepting this water at the edge of the roof is normally the function of gutters and drains.

Rising damp

The threat from rising damp is ever present in temperate climates. Moisture is found in the ground at virtually all times of the year. While the greatest threat may be during wet periods when the moisture content of the soil will be high, the moisture persists in all but the surface layers, even in dry periods. This is even above the water table. If a building is built below the water table, and comparatively few are, then the building is under an additional threat of water pressure assisting entry through the fabric of the building.

The major action associated with rising damp, as the name implies, is a transfer of moisture up into the building against the pull of gravity. The water travels through the fine network of connected pores in permeable materials by capillary action. This rise within the pore system in the materials can reach considerable height above the source of the moisture in the ground.

It is not normally economic to use anything other than permeable materials for the major components below ground. The usual detailing solution is to provide an impermeable barrier known as a damp-proof course (DPC) in walls and damp-proof membrane (DPM) in floors. Identification of the damp pathways and the detailing of damp-proofing at the junction of the ground floor and the external wall of a house are shown in Figure 10.7.

The detailing and care in construction to ensure continuity within and between these damp-proof barriers become extremely important. Given that the major action is capillarity, which by definition exploits the finest pathways, this indicates

Figure 10.7 Pathways and details to resist rising damp.

that any lack of care in detailing or assembly can have a disproportionate effect. Once damp-proofing is breached the materials on the dry side absorb the moisture and transfer of the moisture will continue to internal surfaces.

Condensation

The descriptions so far have been of the possible pathways threatening a dry interior, essentially from outside, through the fabric. This is, however, not the only source of internal dampness. Primary among other sources is condensation. The moisture for condensation does not enter the building as water but as vapour and most is generated within the building through the everyday activities. This not only comes from cooking and bathing but also is generated in the act of breathing, leaving almost all activities as

generators of vapour which, as a gas, is carried away by any process of ventilation occurring in the building.

Vapour is not a threat to the dryness of the environment (although excessive vapour causing high humidity can be uncomfortable); as long as all the vapour can be held in the air it causes no damp problems. There is, however, a limit as to how much vapour can be held in the air dependent on a number of factors, primarily temperature. If vapour increases at constant temperature or temperature decreases with constant vapour the result may be condensation or the appearance of liquid water. This is most commonly seen on the internal surfaces of window glass which will probably have one of the lowest surface temperatures in the room. Another vulnerable area is at the internal corners between wall and ceiling and behind furniture where ventilation has least effect. Here the surface temperature may be higher but so will the vapour concentrations. As these surfaces, unlike the window glass, are permeable the condensation will soak in and the first sign of a problem may be mould growth.

The temperature at which condensation occurs varies depending on the quantity of vapour in the air. The less vapour in the air the lower the temperature at which condensation first occurs. The temperature at which condensation will occur for a given air/vapour mix is called the dew point temperature where the air is said to be 100% saturated or to have a relative humidity of 100%. If the temperature continues to fall (or more vapour tries to enter the air) more condensation will be produced, as the air can be no more than 100% saturated with vapour.

As well as occurring on surfaces, condensation can be a problem within the materials themselves. Most of the materials of which construction is made are not only water permeable but also gas (and therefore vapour) permeable. The two are, however, not directly related. Gas permeability of a material is a different property to water permeability. When air inside the building is warm it is probably holding more vapour than cold air outside. This difference is sufficient to cause vapour from inside to migrate through the construction towards the outside. As it is warm on the inside and cold on the outside, heat will also be flowing out across the construction, the temperature of each layer of material becoming colder towards the outside. This drop in temperature becomes particularly large across any insulation. Insulation material does not normally hold up the vapour being transferred so this may create the conditions for condensation to occur within the construction. Because it happens inside the materials it is known as interstitial condensation.

Water is now present within the construction. If vapour continues to flow and temperatures remain low the quantity of condensate increases. The presence of water within the materials may threaten the durability of the materials and may also change other properties on which the function is depending: principally their insulation and air permeability values. Depending on the

detailing and where the condensation occurs, it may create dampness on the internal wall. This will depend on there being a pathway for the condensate water back towards the inside face of the construction. Similar forces and mechanisms create these pathways, as discussed in penetrating dampness. The water will travel in capillary paths within porous materials or through cracks. Gravity will work on the water (hardly at all on the vapour), draining it through the construction where it may find a pathway in.

If it does not find its way into the building it will, like absorbed rain, have to dry out from the outside surface where wind, rising temperatures and the sun will dry the surface, creating the conditions for a migration of absorbed water to the surface or to evaporate it back to vapour. This may be a slow process and condensate can remain within the construction for considerable periods of time.

Detailing, including the position of the insulating layer along with ventilation and vapour control layers can limit the risk of condensation and this is discussed in the section on the warm environment that follows.

The warm environment

In temperate or cold climates it is natural to see the function of the building as creating and maintaining a warm environment. In warm or hot climates the function may be characterised as providing a cool environment and this will lead to very different strategies and will therefore need a different analysis to the one described here. In hot climates thermal mass and white exteriors are likely to be employed rather than insulation, which is common in temperate climates. This section will focus on the temperate climate winter heating condition although the summer condition will also be considered.

Feeling warm is not just a case of the air temperature. It is dependent on radiant effects, cooling air movement (draughts or breeze) and humidity.

Radiant effects are a process by which heat is lost (or gained) from the body by radiant exchange. This is independent of air temperature and only requires a cold (or hot) surface to be in direct line with the individual (or another object) for heat to be exchanged.

Areas of glazing, particularly single glazing, can produce such cold surfaces in winter and hot surfaces in summer. The rate of radiant exchange is proportional to the temperature difference and this is likely to become large if the glazing itself is directly exposed to sunlight (heat gain) or a clear winter night sky (heat loss). It is likely to be uncomfortable if activities have to be carried out too close to the glazing. Under these circumstances some form of shading, blinds or curtains may have to be used, either externally to keep the surface at air temperature or internally to cut out any direct radiant exposure from the surface to the individual. Placing warm radiators beneath cool windows will also help mediate the radiant exchange experienced by the individual if the temperature of the cool glazing is not too extreme. Working very close to a radiator can also be uncomfortable due to radiant exchange.

Air movement will come from ventilation, where if the rate of airflow is too high a cooling effect will be felt by evaporation from the skin. This will be discussed in the section on ventilation below. High humidity reduces the efficiency of skin-cooling evaporation, making us feel less comfortable in high air temperatures even with high air flow rates.

The function of the passive fabric of the building in maintaining a warm environment within the building is complementary to the function of the active environmental services systems. It is necessary to see any thermal design to be integrated between the fabric, heating and ventilation arrangements of the building. This strategic decision is also concerned with issues of energy use, clean air and running and maintenance costs. Whatever the overall contribution in each case the fabric will fulfil the same function, only the level of performance will vary.

In a temperate climate, the usual initial concern is the loss of heat through the fabric of the building under winter heating conditions. The more heat that is lost through the fabric by transfer and ventilation, the more heat has to be provided by the heating system. This reflects the wider concern associated with energy consumption and running costs. There is also increasing concern with larger commercial building regarding cooling in summer, dealing with high internal heat gains and reducing solar gain so that less cooling is required. The role of the fabric under these conditions is different to the winter heat retention role. This will, for instance, require greater rates of ventilation, the use of shading on the external face and exploiting thermal capacity rather than lower ventilation rates and insulation of the building skin required for winter conditions.

To consider the external skin of the building it is necessary to understand how the construction responds when a temperature difference exists on either side of a wall, roof or floor. The fabric will almost certainly consist of various layers performing different functions, only one of which may specifically be to enhance thermal performance. However, it is the behaviour of the fabric as a whole that needs to be considered.

It is possible to visualise the materials and layers of construction from the point of view of either ease with which they transmit heat or the resistance they offer to the passage of heat. Generally the description that follows will think in terms of layers offering resistance. The function is resisting the passage of heat. Layers with high resistance have materials normally thought of as insulating materials.

The simple view of fabric heat loss

The simplest view of heat loss through the fabric considers the system of inputs and loss to be in equilibrium. It assumes the temperature on each side to be constant and the rate of transfer to have reached 'steady state'. It also assumes that all heat moves directly through the fabric from one surface to the other. The transfer of heat through the fabric is, under these circumstances, dependent on the resistance to heat transfer offered by the layers of the materials chosen, including any cavities, plus the layers of still air held at the surfaces of the construction.

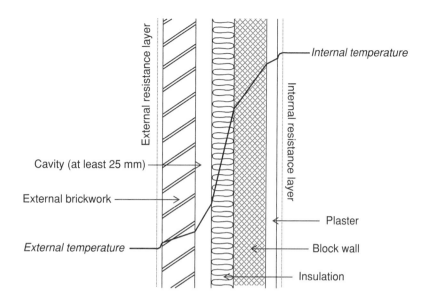

External resistance layer

Internal resistance layer

Internal temperature

Cavity (at least 25 mm)

External brickwork

External temperature

Plaster

Block wall

Insulation

Figure 10.8 Typical temperature drop across insulated cavity wall.

Steady state is the simplest way to visualise the heat flowing through the construction and it illustrates the fundamental influence of the construction. Given a temperature difference on either side of the fabric, heat will flow from the high to the low temperature. The quantity of heat that passes through the fabric, in a given time, the rate of heat flow, will depend on the resistance of the layers including surface air layers and cavities of air. The greater the resistance of the whole construction the less heat passes per unit of time. It is not possible under steady state for more heat to be flowing through one layer than another.

The second influence of having layers with different resistances is the way in which the temperature changes across the construction from the warm to the cool side. This is dependent on the relative resistance of each layer. This is illustrated in Figure 10.8. Layers contributing a large proportion of the resistance to heat flow (the insulation) will experience a large proportion of the temperature drop across the construction as a whole.

Not only is steady state the easiest way to visualise heat flow, but once the resistance of each layer is established it is the simplest condition for which to provide calculations. The rate of heat loss is expressed as a function of the whole construction, combining the effect of all the layers into one figure representing the transmittance of the complete construction. This is a single transmittance value known as the U value.

Heat loss calculated in such a way is, however, based on a description of behaviour that is rarely observed in practice. While the internal temperature may be relatively constant, the external temperature throughout a heating season will vary a great deal. Deciding on a design condition for external temperature becomes crucial to the value of the estimates provided by such calculations. Chosen too low and heating systems will be over-designed for most of the heating season, while chosen too high and conditions of comfort may not be achieved for a large proportion of the days when heating is required. Current design practice in the UK is to take 0°C as the external design temperature.

The accuracy of any estimates calculated in this way is sensitive not only to the choice of design conditions but also to detailing. The assumption was made that all heat moves directly through the fabric from one surface to the other. If the resisting layers remain continuous over a large area this assumption is reasonable. However, whenever the layers are interrupted by structure or openings this assumption must be questioned.

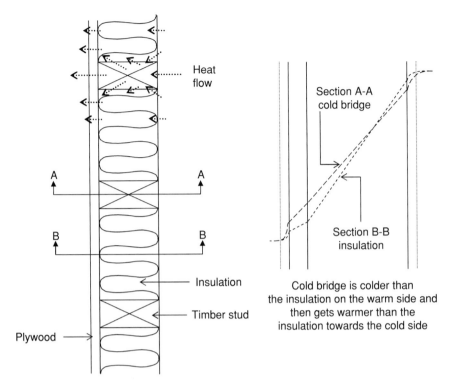

Figure 10.9   Effect of studs as cold bridge in timber framed wall panel.

Cold bridging

Figure 10.9 illustrates the effect on heat flow through construction where the studs in a timber-framed wall panel do not provide a single layer. The studs are said to be providing a cold bridge by the diversion of the heat so it does not pass directly from one side to the other. This will also occur at junctions of floors and around window openings. Heat will flow naturally towards the lower temperature, distorting the temperature profile. This thermal bridging has the effect of losing more than a simple heat loss calculation would suggest. Hence where a building has little continuous high resisting fabric with frequent penetrations or openings, the detailing to reduce this effect needs to be considered. The potential bridge at the head of a window and the detailing to minimise the effect are shown in Figure 10.10.

Non-steady state heat exchange, capacity and diffusivity

While a building is being heated the internal temperature will be relatively constant. The outside temperature will vary much more, even on a daily cycle. Not only does this involve the air temperature but also surface temperatures will be directly affected by radiant exchange by direct exposure to the sun (gain) and to clear night-time skies (loss). Surfaces, particularly if dark in colour, will absorb heat if directly exposed to the sun and lose heat if directly exposed to a clear winter night sky. Surface temperatures can by this mechanism reach values significantly different to air temperatures.

Buildings may not be continuously heated. Intermittent heating may be necessary to match the occupation pattern. Under these circumstances both internal and external temperatures

Figure 10.10   Cold bridge and details to reduce heat loss at window head.

are changing, often in the opposite direction, with internal temperatures rising just as external temperatures are falling.

In order to visualise this dynamic thermal behaviour it is necessary to think about how rapidly the temperature profile in the fabric changes in response to temperature changes at the surface. The thermal properties of the materials and thickness of each layer will determine the speed with which the fabric will respond. This needs not only a measure of resistance or insulation value of the layers but also one of thermal capacity. Capacity is a measure of the amount of heat necessary to create a given rise in the temperature of the material (or for the heat to be given up during an equivalent fall in temperature). The rate at which the temperature within the layers changes will be quick when there is little resistance to heat flow and little thermal capacity. With increasing resistance or capacity of the layers, changes in temperature within the materials will be slower. This measure of the speed at which the internal temperature-profile changes is a function of both resistance and capacity and is known as thermal diffusivity. Generally, materials offering high resistance are low density with large volumes of air entrapped in the materials, while high thermal capacity materials have higher density and hence a lower resistance to heat flow.

Analysis of response and capacity becomes more important than insulation in climates where the large changes occur between night and day, such as in desert areas, rather than between summer and winter, as in temperate climates. Hence, heavier, thermally massive materials are used. This is similar to the summer cooling designs in temperate climates where the fabric is used to store the heat generated during the day which is then purged at night.

Analysis of response is also important where buildings have to be heated for short periods, when thermally light building may be appropriate. Another strategy would be to have a heating system that is chosen with radiant heaters that heat the individual directly without heating the air so that the fabric has little role to play in feeling warm other than reducing draughts.

Position of the insulation

When only considering heat loss and taking the simple view of steady state conditions, the position of the high-resisting, insulation layer is of no consequence. When the behaviour of the fabric is being analysed for any purpose where temperature profile or response time is important, the position of the relative resistance offered by each layer may become significant.

Figure 10.11   Temperature gradient with position of insulation in cavity wall.

If the insulating layers are near to the inside of the building (warm side) then the outside layers will have a smaller temperature rise in order to achieve steady state. In these circumstances not only is steady state approached relatively quickly but also there is little heat stored in the fabric even if constructed with relatively high thermal capacity materials.

Often more significant is that if the insulating layer is on the warm side then there is an increased risk of condensation with the lower temperatures that arise on the cold side of the construction. Figure 10.11 shows the temperature profile for walls with the insulating layer in different positions. The converse is, therefore, that insulation on the cold side reduces condensation risk but may increase warm-up periods where intermittent heating is required.

If the condensation risk is high then additional detailing will be required. This will normally take the form of either a vapour control layer or the ventilation of cavities to take away the vapour migrating across the construction. Ventilation can be particularly effective in roof construction.

Ventilation

Maintaining a warm environment is affected not only by the choice and detailing of the fabric but also by the amount of ventilation. The direct loss of warm air from a room under winter conditions accounts for a significant proportion of the heat loss. Ventilation involves air exchange

and therefore airflow. If the speed of this airflow becomes too great it might be called a draught, increasing heat loss from the body so we no longer accept the environment as being warm even if the air temperature has not changed. Under summer conditions the increased speed of airflow may be called a breeze and is welcomed in order to maintain a comfortable thermal environment. In temperate climates, ventilation is the main means of maintaining a comfortable thermal environment under summer conditions. Air conditioning, being expensive to install and having high running costs, is restricted to a limited range of buildings in these climatic zones.

Ventilation, however, is not just a consideration for maintaining a thermal environment. It also plays its part in two other functions. First, it removes water vapour which reduces the risk of condensation and hence dampness in the building. Second, it removes stale air, maintaining a clean environment, keeping down the level of air-borne pollutants and odours to a safe and comfortable level. This will be covered in the section on the clean environment for there is no doubt that an exchange of air is necessary within buildings and some means of achieving a controlled level of ventilation has to be considered.

For ventilation to take place it is necessary to have two conditions. There needs to be a pressure difference to move the air and the presence of openings to let air in and let air out. This assumes a supply of air that is suitable to replace the air in the building. If this is not the case then

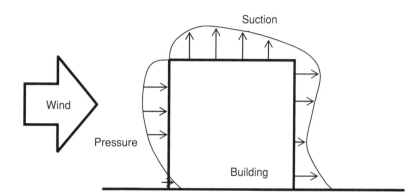

Figure 10.12 Typical wind pressure around buildings.

cleaning by filters or dehumidifying may be necessary, but this will not be considered here.

The need for a pressure difference and openings is true for both natural and mechanical systems. The way they are provided and the means of control are, however, very different. They may be hybrid approaches, the simplest of which is mechanical extraction to increase (or at least control) the pressure difference in order to dictate the exit opening but leaving the fabric to provide the inlet openings. Even with a full mechanical system, the fabric must be detailed and built so it is not too leaky as this could influence the performance of the mechanical system.

The part played by the fabric in the provision of ventilation is one of providing openings. These are not only those specified for this purpose such as air bricks or chimneys but also the 'cracks' that are present around opening windows and doors. Sealing these cracks with 'draught excluders' is often seen as a cost-effective way of saving heat. This may be true, but it may also exclude a level of ventilation necessary to maintain a dry and healthy environment. Windows with draught seals should be provided with some other ventilation process such as trickle vents if there is a danger of low ventilation rates in the building.

Pressure differences created by the wind occur naturally around buildings. Typical pressure distributions around a building are shown in Figure 10.12. Air will enter any openings on the side of the building experiencing positive pressure and escape from the side with negative pressure.

A second process that can be harnessed to create natural ventilation is the stack effect. This does not induce a cross flow of air but a vertical air movement. It relies on a temperature difference with warm air at the bottom and cooler air above, but it will be assisted by wind if the upper opening is in a negative pressure and the lower opening in a positive pressure region on the outside of the building.

Since the primary force inducing natural ventilation is the wind, it is clear that control has to use openings in the fabric and that this control will have to account for a variety of conditions experienced locally by the building. Local conditions can be highly modified by adjacent buildings which may have a profound effect on the levels of ventilation achieved in practice.

Visualising these processes demands an analysis of the wind pressure and temperature differentials as well as the pattern of opening provided by the fabric. These can be highly complex and particular to an individual building. Mathematical modelling, wind tunnel and salt bath trials can be used to inform this process of visualisation for building using natural ventilation. This can be a major strategy in this important area of the building's performance as it can greatly reduce the energy use in a building.

The design of the building as a whole has to be considered in order to harness the forces in natural ventilation. One way this can be achieved is with the use of the atrium or street (generally

Figure 10.13 Natural ventilation using atria or internal street.

an atrium along the length of the building). The basic design and the natural ventilation forces are shown in Figure 10.13. This design can also take full advantage of natural lighting, again reducing energy use throughout the life of the building.

The light environment

To fulfil the majority of its functions the enclosure of a building will almost certainly use materials that do not transmit light. This means that if the user is to make use of natural light, specific consideration will have to be given to getting light into the building. Since most buildings are occupied during the hours of both daylight and night-time, provision has to be made for artificial lighting as well as natural lighting. These two may be separate in that the building is to be designed to operate during the day without the use of artificial light. Alternatively there may be an element of permanent artificial supplementary lighting for daytime activity. This does have considerable running cost and energy implications as well as considerations for the consistency of level and quality of lighting provided. One further issue in lighting is that, over and above providing general levels of light, there may be a need for the provision of task lighting or dramatic or mood illumination, which may also be provided by either artificial or natural lighting depending on circumstances.

Natural lighting

The simplest way to allow light into a building is to leave empty openings in the external enclosure, but this is unlikely to be satisfactory given the other functions of the walls and roof. The materials available for enclosure that transmit light are limited, being primarily glass and plastics. Increasing thickness of these glazing materials reduces the amount of light transmitted, limiting them to relatively thin sheet applications.

These materials will have limited strength and are therefore unlikely to be able to carry structural loads from the building, although they will have to carry wind loads to the structure. They will most often be framed to provide the strength required. As single sheets, the glazing materials will transmit heat easily, giving high heat loss in winter and high gain in summer and will provide limited sound resistance. At the same time they must remain waterproof and are often used to control ventilation by providing sections which can be opened. They may have limited ability to provide security and privacy.

These components, such as windows and roof lights, become sub-elements of walls and roofs of which they form a part. They have to contribute to the performance of the element as a whole, but it has to be accepted that they will have different performance levels. These components may also be made of very different materials to the fabric that surrounds them and the detailing of the fixings and jointing to the main enclosure materials has to be thought through independently and yet be complementary with the consideration of the elements as a whole.

In addition to all these considerations the size and shape of windows has a significant effect on the appearance of a building. They are an integral part of most styles of architecture, often playing a major part in giving the style its distinctive character.

Against this background it is clear that the solution to providing natural light may often be a compromise with the other functions of the building. Its own importance in function and design terms varies greatly depending on the use of the building and the designer's interpretation of its importance to the design. The use of multi-layers of glass or plastic components is now common and recent development in materials has given these multi-layer transparent components improved thermal and sound performance and even strength and fire-enhanced properties to help security functions.

What is clear is that the energy use in electrical artificial lighting is high and therefore the provision of high quality natural light can play a large part in achieving sustainable development.

Source of natural lighting

Being dependent on the sun, the quality, intensity and time during the day when natural light is available are highly variable. These vary with latitude and with orientation. In the northern hemisphere light from the south is generally brighter than from the north. It does, however, vary in intensity, often causing uncomfortable or disabling glare rather than even illumination. Light from the north is more consistent and reliable, colder and really dramatic. This made it suitable for factories and warehouses in the characteristic saw-tooth roof-line of north lights. For other purposes, variety and qualities of brightness and warmth may be more appropriate, which, with some control for glare, is best provided from the south.

Natural light also varies with the time of year and the weather conditions. The hours of daylight vary with the seasons, and the light levels and the angles from which the sun can be directly viewed will vary. This will give a quality of light that varies with the seasons but also means that shading may have to be variable to match the angle and direction of the sun. The cloud cover affects the quality of the light. The highest level of light is available with no clouds

covering the sun whereas storm clouds can all but eliminate the light, creating near-night conditions during the day.

With all this variety it is unlikely that performance levels can be assured throughout the year. Bright light may need to be shaded, while storm clouds may dictate that artificial light is needed during the day. This variety is greater in temperate rather than in hotter climates where the sun is more predictable and remains higher in the sky. In buildings designed for these hot climates windows may be small and/or be permanently shaded by shutters or verandas facing the main path of the sun.

Variety of natural light is not necessarily undesirable. Changing levels of intensity and colour give information about outside conditions. They keep occupants 'in touch' with natural conditions, marking the passing of the natural cycles of the day and the seasons. In addition the light received brings with it a view from outside. Windows, and perhaps to a lesser extent roof lights, allow occupants to look outside to see what is going on. In many cases, landscape (or cityscape) can be designed to provide spaces seen as visual extensions to the internal space. This gives vistas and panoramas that enhance the activity for which the building is designed. The increased cost of offices on upper floors of skyscrapers and penthouse suites is based in some measure on the view they provide.

Despite the technical difficulties of providing the performance of the enclosure where lighting openings are to be included, and the probable additional capital and running costs, windows and roof lights are judged to be good value (if not essential) in the provision of views, light and the influence they have on the appearance of the building. This makes the technical challenges particularly acute.

Penetration of natural light to internal spaces

Given the prevailing external lighting conditions, the quality and quantity of light reaching the internal space are controlled by the size and shape of the openings, the opacity of the glazing material and the shading or other control mechanisms provided. The proportion of light then available penetrating into the space depends upon area and shape of the room, particularly its depth. Lighting levels reduce with increasing distance from the windows, but penetration can be enhanced by careful choice of shape and position of the openings.

Lighting levels from natural light are also influenced by reflections from objects or surfaces in the room. Surfaces that are shiny or light in colour will reflect a greater amount of light, raising the level of light at the working surface. In the same way as it is possible to provide shade from the direct glare of the sun at the window it is possible to provide reflectors to enhance the levels of light incident at working positions within the room.

Predicting lighting levels from natural lighting is now well established. Data on the external conditions and calculations that take window shape, depth of room and the reflection of surfaces into account are available to predict the level and reliability of natural lighting into a building.

Artificial lighting

For any building that is to be in use after dark the need for artificial lighting is clear. The provision of this lighting will assume that there is no light available from outside. Windows left uncovered will then act to transmit the internally generated light out and the view will then be available from the outside of the inside, affecting privacy and security. Heat loss will also increase as the temperature falls at nighttime. Given these circumstances, curtains or blinds become associated with window design.

After dark, not only is lighting for internal spaces lost, but also the exterior of the building loses its illumination. This has implications for its appearance and for security and safety. These two may require different types of lighting. They may need different levels of lighting, colour and modelling requirements. However these are resolved there is often a need to provide sources of artificial lighting outside as well as inside the building.

If levels of natural lighting are assessed to be inadequate in the daytime then artificial lighting may be seen as a solution even during the day. This is particularly a problem in deep building where the middle of the space is some distance from the external walls. In single-storey buildings (or the top floor of a multi-storey building) this can be solved with roof lighting. In very deep multi-storey building the use of light wells or atria can bring light to a central space, making light available to areas some distance from the outward-facing walls. In other circumstances, artificial lighting will have to be used, although this has energy and running cost implications over the lifetime of the building.

For many buildings the daytime use of artificial lighting is only associated with limited periods when the external conditions are poor. In these cases the lighting designed for night-time use may be switched on. However, even if general lighting levels with natural lighting are satisfactory, specific task lighting may still be required and provided by artificial sources.

Sources and control of artificial lighting

The source of natural light is the sun, often obscured by cloud, so its quantity (level of illumination) and quality (colour, warmth, etc.) are variable dependent upon the time of day and the seasons of the year. With artificial light sources, the parameters of quantity and quality can be chosen. Since the case for artificial lighting also has to consider energy consumption, it is important to be able to identify the energy use for a given output since this also varies greatly between the different sources.

The demands made for lighting both internally and externally vary greatly and hence a wide range of sources has been developed. Originally powered by oil or gas, artificial lighting is now almost exclusively powered by electricity. The initial characteristic of the light is governed by the way the electrical power is converted into light. Whether the light is generated by the incandescence of a solid material (the light bulb) or the excitation of a gas, it is this conversion process that determines the basic colour and the power consumption of the source.

This light can then be harnessed to best effect by the fitting and positioning of mounted units. Directional spot lights get their power and colour from the light source, but their directional capacity comes from the way they are housed and even focused, with lenses to minimise scatter as with theatrical lighting. Colour can also be added to white light with filters. For general lighting, the fluorescent tube, mounted in a reflective box, or luminaires with a diffuser on the front to give an even light with limited glare and then mounted in the ceiling at relatively regular spacing provide an overall level of illumination with some control over colour with the choice of tube.

More general control over individual rooms or sections of large areas is achieved by switching the electricity on or off or dimming to reduce the level of light emitted. In many situations this control can be manual by providing switches for the occupants to operate. For energy saving some level of automation should be considered. Often based on movement sensors to identify occupants, automatic switching will turn off lights when rooms are not being used and turn them on again when people return. This may have to be set to leave low levels of illumination if safety might be compromised. Light level sensors can be used to switch on lights as levels of natural lighting reduce. It is particularly effective to bring in banks of lighting parallel to the windows to supplement the natural lighting at the back of the room without using lights close to the windows where natural lighting is still satisfactory.

As with natural sources, artificial light received at a point in a room is not just in a direct line from the source but also arrives from reflections from surfaces. The strength of these reflections is dependent on how close the surface is to the source and on the texture and colour of the surface itself. This can be used either to increase the level of illumination (quantity) or to create effect (quality) when used in conjunction with the source, its housing and its mounting position.

The predictive methods and the data available for the sources concerning the levels of general illumination are well developed. Both predictive models and data on reflective surfaces and the manufacturers' values for the sources and the luminaires are freely available and reliable. Even predictions of loss of efficiency with time, energy usage and heat generation are possible with the current data available. Architectural lighting and effect or mood lighting are less susceptible to calculation and need a measure of imagination, but some calculation and computer modelling can give clues to comparative effects especially in terms of power and the interpretation of light and shade for soft or dramatic effects.

The acoustic environment

Sound can be considered either as wanted or as unwanted noise. Sound that is wanted can be enhanced or degraded by the fabric of the building particularly by the finishing and the shape of the room in which the listener is located. Most wanted sound is generated within the space occupied by the listener, most often associated with talking or music. Some wanted sound comes from outside the room. Often sounds from the natural environment need not be totally eliminated since they bring an awareness of the wider world. Like a view or the changing quality of natural light from a window they limit any sense of isolation that can be psychologically undesirable. Very low background noise conditions the ear so it picks up small noises that become intrusive. The lack of any external sound can also be disturbing as sound is used to identify threats. In modern building the possibility of natural threats is now supplemented by a number of alarm systems provided within the building and it is necessary for these to be heard by all the occupants. Fire and burglary are the most obvious and these must be audible in all parts of the building when activated.

An increasing number of modern sounds are now considered as unwanted. They are categorised as pollution and as such are now being

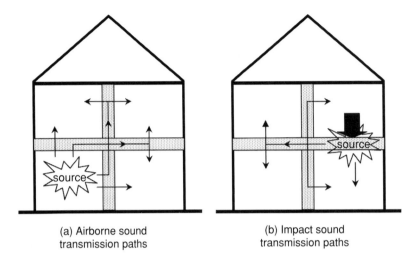

(a) Airborne sound
transmission paths

(b) Impact sound
transmission paths

Figure 10.14   Sound source
and transmission paths.

given much attention in reduction at source, but this still often leaves excessive noise that has to be reduced through careful choice of materials and detailing of the construction.

Characteristics of sound

Sound has two characteristics as identified by the ear: frequency (pitch) and loudness. Frequency determines how high or low we perceive the sound to be. If too high or low it becomes inaudible to humans, although some pets can hear higher frequencies, and some low frequency sound does affect other organs in the body and causes discomfort, but these sounds are very rare. Combinations of frequencies can be harmonious, as in music, or carry information, as with speech. Groups of frequencies that are disharmonies or carry no information are often called noise and, while useful as alarm signals, are generally considered undesirable.

Loudness is associated with 'volume'. Because it has energy or power it has the capacity to cause pain if it excites the mechanism in the ear to excessive movement. At the other end of the scale, quiet noise can be irritating since it is frustrating if you think it is carrying information that you cannot obtain; or it is continuous and begins to intrude on your consciousness.

All sound, whether wanted or unwanted, reaches the ear through vibrations set up in the air around us. These vibrations from an airborne sound can also travel through the construction and hence the provision of a barrier will only modify and not eliminate noise. Further, impact sound created by striking the construction itself, such as footsteps on a floor, will create a sound not only within the room but also transmitted to other areas by the direct connection of the structure and fabric of the construction. The transmission of the sound around the walls and floors directly between source and listener is by means of flanking paths. These noise sources and transmission paths are shown in Figure 10.14. In addition, mechanical plant has the potential to create noise and this can be passed through the building via the services duct and pipework systems.

Sound created within the space

The sound will travel out from the source and there will be a natural decay of its power as it moves away. If in a direct line with the source, the listener will receive the sound directly. The sound will continue through the air in the room, striking other objects and eventually the floor, walls and ceiling that make up the room. Sound reaching openings in the floor, walls and ceilings will pass straight through. The sound that strikes objects or the surfaces in the room will be partly absorbed and partly reflected depending on the nature of the surface itself. The reflected component will return into the room again, where it may be received with a reduction in power and fractionally later than receiving sound direct from the source. These reflections will continue to be heard by the listener until the power in the reflection is too low to be audible. The quality of the sound is, therefore, highly dependent on the acoustic (absorption and reflection) characteristic of the objects in the room and the surfaces. However, since the sound loses power with distance, and takes time to return as a reflection, the size and shape of the room are also important. The period of time each individual sound takes to decay to become inaudible is known as the reverberation time. With highly reflective surfaces in a large room reverberation times will be increased.

For both speech and music it is important to have a quiet background; little or no noise from outside the room and little or no generation of sound within the room other than either the speech or music. For both it is necessary to have a good strong direct path from source to listener, but for reflections and reverberation time the requirements for speech and music are very different. For speech a strong early reflection reinforces the sound, but later reflections interfere with clarity of the separate sounds. Reverberations as low as a twentieth of a second are required. For music it is the later reflections that enhance the sound, enveloping the listener in the sound field. Now reverberations of one to two seconds are required. This enveloping also requires the room to have a certain volume to maintain a smooth and prolonged decay of the sound.

There is one other phenomenon that can occur if a sustained sound with a dominant low frequency is coincident with the dimensions of the room. This is most likely to arise from vibrations from outside the room, which will be considered later, but the characteristic is associated with the room itself. If the room dimension is a function of the frequency (wavelength) of the sound, a standing wave may be generated sustaining that frequency.

It is clear that the major influence the construction can have over the quality of the room acoustics is associated with the reflection/absorption of the finishes, having taken into account the influence, particularly absorption, of the objects (including the listener) in the room. In general terms, hard, flat surfaces are highly reflective. As the surface becomes progressively softer and more textured the more absorption is increased. Soft, fissured or even re-entrant surfaces are sound absorbent as the energy in the vibrating air becomes dissipated within the cavities of the irregular surface.

Airborne sound made outside the space

Increasingly airborne sound is unwanted, but, as has already been pointed out, total sound isolation may not be a sensible performance requirement. The sound will move out from the source as before. If it is in the open air it will be modified by wind and other atmospheric disturbances, but no such disturbance would be experienced if the sound were created in another room. The sound is again going to be influenced by objects it meets, being either absorbed or deflected, but at some point it will impinge on the dividing construction such as walls and floors.

An opening in the construction will not change the sound; it will enter (or leave) the room unaffected. It is important to appreciate the significance of the size of openings in transmitting sound. Small openings have a large effect, considerably reducing the sound reduction performance of a barrier, however well detailed.

Where the sound meets the walls and floors of the building they will act as a barrier. The construction will reduce the loudness but not equally for all frequencies. The action of the construction in reducing the noise is two-fold: it 'dampens' the vibrations due mainly to the mass of the construction and it isolates the vibrations if discontinuities (layers of resilient materials or cavities) can be introduced in the detailing. Hence the overall performance of a wall, floor or roof in attenuating sound will depend on the openings, mass and discontinuities and any sound-absorbent materials in the cavities. Sound

Figure 10.15   Timber joisted floor and details to provide soundproofing.

will also be carried deeper into the building by other elements attached to the initial barrier along flanking paths, making edge detailing particularly important. Figure 10.15 shows possible soundproofing details for a timber-joisted floor, illustrating how mass, discontinuities including abortion in cavities and edge details can be integrated into a basic structural floor.

Care has to be exercised in assessing the performance of cavities. If the cavity is too small it has little effect as the vibration energy is transferred directly across the cavity. This is known as acoustic coupling. It happens in standard double-glazing where the gap between the glass panes is small and therefore the soundproofing effect is almost entirely due to the additional mass of the second pane of glass. For cavities to be effective they have to be approaching 200 mm or larger.

While the potential sound reduction lies in the basic construction of the floor or wall the part played by edge conditions and flanking paths means that the actual transmission loss when the building is in use may be less. Transmission loss of the proposed construction can be measured in the laboratory to provide a sound reduction index R_w normally between the frequencies of 30 and 315 Hz. A value of 40 dB reduction will render loud talking inaudible. This value helps to select appropriate specification, but these laboratory-based values may not be achieved when incorporated into the final construction on site, depending on the detailing and any variability arising from the site operations. It is possible to measure on-site performance, where the transmission loss R'_w will be quoted.

Sound generated by striking the fabric or from mechanical services

Sound is created by directly setting up vibrations in the construction itself. These vibrations will be much stronger than those set up by airborne sound reaching the fabric of the building. Flanking paths created by continuity between elements of structure and fabric are important since if the vibrations can travel across junctions between elements and via joints and fixings the noise will penetrate deeper into the building. Mass will again play its part in damping the sound, but rigidity and natural vibration or resonance of the components sympathetic to the frequency of the noise may become important. Discontinuities and edge detailing are particularly effective in limiting the transmissions of these sounds.

Striking hard surfaces creates the greatest sound, not only in the room but also in setting up vibrations throughout the structure. Soft surfaces dampen the original impact and absorb the airborne sound that results.

There are particular problems associated with the pumps and fans of the services. The continuous rotation generates noise, but the movement of the rotation may also set up continuous vibrations in the construction to which it is attached. This may be the fabric of the building supporting the pump or fan or it may be the pipe and ductwork to which it is attached. Isolation or discontinuities have to be used to mount the pumps and fans and connect them to the distribution system.

Methods of prediction

Definitions of a comfortable acoustic environment are well established for a wide range of activity including spaces with both general and specific acoustic demands, e.g. offices and concert halls. There is a great deal of empirical measurement on the characteristics of the materials commonly used in construction, and predictive calculations based on theory can utilise these values to predict performance in a wide variety of circumstances. Modelling, originally physical but increasingly on the computer, can be used in design of specific spaces.

For the predictions to be fulfilled much care is necessary in the design detailing and the production phase. Identifying isolation discontinuities and in particular edge fixings to elements of the construction are of prime importance. During site operations, fixings and edge details need particular care. Gaps at junctions and the rigidity of connections can be crucial in achieving acoustic performance.

The clean environment

As with all aspects of environmental performance it is necessary to be clear what activities are being carried out in which part of the building. Acceptable standards of cleanliness will have to be defined for the occupants but may also have to be considered for the materials and equipment to be housed and used within the building. For all these aspects of the activity, cleanliness will have to be defined against two complementary sets of criteria: those that would be detrimental to health or operational requirements and conditions that would be socially defined as unacceptable.

The socially defined criteria for cleanliness may not be detrimental to health or operational efficiency. They will, however, contribute to the social and psychological health of the occupants. The distinction between health of occupants and the efficiency of operations is also important. Definitions of cleanliness will have to be identified against ideals for hygiene, the control of pathogens harmful to life or health. They will also have to be defined for the materials stored and equipment operated in the building. Levels of dust and dirt may not be detrimental to life but may interfere with the operation of equipment.

Another way of considering cleanliness is to identify the possible undesirable components of the environment as physical, biological or chemical. The term 'clean' is most associated with the absence of dust or dirt. These particles will be of various sizes and possible composition. They may be inert minerals but are also likely to include matter of biological origins (dead skin cells). They may also be chemically reactive if associated with, say, industrial waste. Particles are most commonly associated with surfaces but can be suspended in the air. The air, however, also carries other potentially undesirable components. Gases can mix with the air and odours can be carried in the air. The source of both gases and odours can be either natural or manufactured and while some can be a health hazard, some are only a nuisance. Some gases, such as carbon monoxide produced from faulty heaters, or burning of fossil fuels have an immediate deadly effect, while the naturally occurring gas of radon from some ground conditions only shows its effect after many years. Other gases such as methane and carbon dioxide from the degradation processes in the ground such as landfill sites that are now being used for development are not just a health hazard but also introduce the safety risk of explosion. In addition to particles, gas and odours, it is now possible to detect appreciable amounts of electromagnetic flux, the long-term health effects of which are still under investigation.

It is clear that surfaces and the air are two major areas of concern in providing a clean environment. Water is also a medium that has to be clean. Once it has entered the building at an appropriate quality, avoiding cross-contamination becomes a priority in the design of the water distribution systems.

Surfaces, air and water are all open to inspection. Occupants and equipment come into contact with them. However, within the construction or in unused or not often visited spaces there is the possibility of infestation. Varying in size from cracks in the construction to whole loft spaces, there exists the potential for infestations of animal and plant life. Infestations of insects, rodents or birds are all possible, as are bacteria in some locations. Plant life is most likely to be mould or other fungal growth, as these often do not need light to maintain growth.

Infestations of this type not only are undesirable against the notion of a clean building but also can be harmful to the fabric of the building itself. Control of these infestations is important for the durability of the construction as well as maintaining a clean environment for the occupants and the operations of the building.

It is, therefore, necessary to undertake an analysis of the air in the building, the surfaces of the construction visible and in contact with the operations in the building and the infestation potential of the construction and unused spaces in the building.

Clean air

It has long been recognised that particles in the air, particularly from smoke, result in a high incidence of respiratory disease as well as making surfaces, including clothes and skin, dirty. Similarly it is recognised that dampness in the air is not desirable and that limits to humidity are sensible. Although not as much of a problem as foreseen by early medical predictions, some disease is carried in the air, although often associated with water droplets suspended in the air, and breathed in with the air itself. Smell is another component of the air that, if allowed to concentrate, can become unpleasant, although it is not really a health hazard in itself. It is now being recognised that very low levels of chemical and ion (radiation) concentrations in the air can also be injurious to long-term health or have a low-level debilitating influence. The full implications of many of these are not yet understood and hence safe levels have not yet been established. It appears, however, that as we 'clean' our environment we improve health but then become susceptible to thresholds of other components in the environment that then redefine our levels of performance.

This general process of demanding new levels of cleanliness has also been true where the activity in the building has been manufacturing. This can be seen at its most obvious in the electronics industry where some processes are now housed in areas called 'clean' rooms.

Clean air is maintained primarily through ventilation. Most harmful components carried by the air are generated from within the building itself and the process is one of replacing the contaminated air with 'fresh' air from outside. Increasingly the air outside also contains contaminants and, therefore, some measure of processing of either the stale internal air or the incoming outside air has to be undertaken. Typically this is the process of filtration with a variety of processes used for physical, chemical and even biological cleansing.

The process of ventilation, both natural and mechanical, has been described earlier in this chapter. The necessity of a pressure difference and openings to create the air movement is the same, but the rates of ventilation may be significantly different. It is necessary to identify not only the types of contaminants but also the rates at which they build up and the threshold levels at which the air is to be considered no longer 'clean'. Like all aspects of the clean environment the performance level of clean is not absolute. There are only acceptable levels or thresholds of contaminants, many of which cannot be detected by sight or smell and require monitoring in other ways. Traditional methods of leaving the occupants to open windows when the room becomes 'stuffy' will still be satisfactory in many situations. However, contaminants such as radon are tasteless and odourless and therefore have to be detected by other means. As a gas from the ground, its control will require gas-proof layers similar to the damp-proofing discussed earlier in this chapter.

Cleanable surfaces

The specification of surfaces is not that they shall be clean, for all new materials are clean. The requirement is that they shall be cleanable. This demands that the nature of contaminants must be known, as must the means that will be used to clean them. It is possible to speculate that the widespread use of fitted carpets has been dependent on the invention and affordability of the vacuum cleaner.

Taking the lifetime cost to the user of a surface, it is more than likely that more will be spent on cleaning a surface than in its initial purchase. In the domestic situation, reducing the time needed for cleaning may be seen as quality of life; to a commercial undertaking this will inevitably mean money.

The method of cleaning will be dependent on the contaminant. It is possible that some cleaning agents may be aggressive to some materials and these will have to be excluded from any possible specifications. Much cleaning associated with hygiene will be water based and hence impervious surfaces will have to be specified. The specification of surfaces with respect to cleanliness will only fail if they cannot be cleaned

in the manner prescribed. Lack of cleaning in use does not constitute a failure of the specification.

Infestation

Many infestations are unpleasant rather than a threat to health. Infestations occur where either plant or animal life breeds and lives within the fabric or within voids created by the detailing or within the furniture and stored items, particularly food. The scale of infestation and the size of the animals vary considerably as does any definition of what might be unacceptable. Generally, in the developed world, very low levels of infestation are tolerated. Performance specifications would wish to exclude the larger animals such as birds, mice and rats and most forms of insects. Some life sharing a building is, however, inevitable. What is shelter to us provides shelter to many other life forms unless the choice of material and detailing preclude them.

The first line of defence with animals (including insects) is exclusion of the mature adult. Although this is possible with animals and larger insects it is often not possible with the smaller species and bacteria. In the case of plant life it is the seeds or spores that have to be excluded and this would prove very difficult. Openings that lead to internal, dry cavities have to be detailed and constructed with care to ensure minimum dimensions to exclude adults. Larger openings have to be protected with screens where infestation is possible.

If exclusion is not an option the key to stopping infestation is to understand the life cycle of the potential infester and the conditions necessary to sustain each stage of its life. Removing food sources may deter an infestation, but for most infesters it is only shelter and protection that they gain from the use of the building which allows them to breed and hence establish an infestation. If it is difficult to limit the conditions suitable for habitation, positively injurious conditions to the health of the infester may have to be created. While this may be part of the specification, as with wood preservatives, it is more likely to be arranged by the user during occupation of the building.

So far, infestations and breeding patterns have been considered. In some cases it is necessary to exclude adults from the actual rooms, not because they would cause an infestation but because they would be a nuisance or health hazard in themselves. This would include flying insects, either disease carrying or able to inflict injury. Similar detailing is necessary, although this may be contrary to requirements for ventilation.

Our knowledge of the animals and plants that are likely to seek shelter within our building is well developed and much traditional detailing takes the elimination of infestation into account. However, we have to be ever vigilant in that poor detailing or poor construction may lead to colonisation and infestation.

The safe environment

The need for safety in a building arises from the perception of a threat. Understanding the nature of the threat determines the performance required of the construction, the nature of failure and hence the choice of materials and detailing. Threats can be seen as arising from natural events, fire and from within society, where they are associated with security.

Natural threats

Many of the natural threats have been discussed in other sections of the text. For example, infestation and the presence of landfill gases such as methane are discussed in the section on maintaining a clean environment. They were seen as a direct threat to health rather than to safety, although clearly some safety measures are designed to prevent injury and damage to health. This distinction is arguable, but this text has chosen to discuss these health issues with the general requirements for a clean environment. Irrespective of how the distinction is made, the choice of materials and detailing will be the same, and these were discussed in the previous section so will not be covered again.

Natural threats not yet discussed come from what is often termed an 'act of God'. It includes the natural extremes of climate, such as high winds and excessive snowfall, and other natural occurrences such as earthquakes. This threat can be identified with the performance of stability and strength. These natural threats exert forces that the structure and fabric have to withstand without failure. Processes of structural analysis and design are now well established even though the natural forces involved are often complex and dynamic. The dynamic nature of the forces means the solutions may have to be carefully worked out in terms of the response of the structure and fabric of the building. This is discussed in Chapter 11

An extreme of climate is flooding. Practical steps that can be taken to hold back water inundation in the choice of construction fabric are limited. Infrastructure projects in coastal and river defences for low-lying areas may limit this threat. Often location and the initial choice of levels for the development are the most effective measures to take against this threat.

The threat of fire

Fire can be a natural threat, but its probability is much higher in the built environment than in the natural environment. The source of most fires in building comes from the activity within the building or a malfunction of the services, particularly electrical. It is also associated with the possible threats from society itself when associated with arson. However it starts, it is seen as a major safety issue in the built environment and its analysis concerns the whole building. It has an influence on most aspects of the construction. There is now much regulation concerning fire safety design and this shows the level of threat that is perceived in modern buildings. Fire is a major threat to life, but there is also a financial implication in the destruction of the property and therefore the design of a building often takes both into account.

Many of the requirements for construction are now laid down in codes, standards and regulations. When operating with any highly codified rules it may be difficult to apply them to new design ideas. There becomes a need to return to

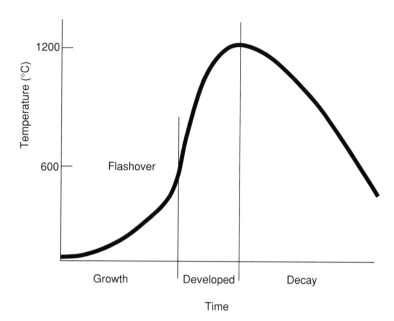

Figure 10.16 Typical fire time–temperature curve.

an engineering approach. This has always been difficult given the complex environment of a fire, but computerised modelling can help in the evaluation.

Whether making choices by using published standards or by fire engineering techniques there is a need to understand the progress of a fire in a building in order to make informed choices.

Like any environmental design, it is necessary first to understand the conditions in which the construction has to perform. The progress of a fire in an enclosed space follows a fairly consistent pattern. This is shown in the typical fire curve illustrated in Figure 10.16.

For a fire to start at all there has to be combustible material (fuel), oxygen and an ignition source. Fire in a building is really a series of fires starting at various time intervals in each room and each following the basic development on the fire curve. Fire in each space requires ignition and then the amount and type of fuel and the rate of supply of oxygen will determine its progress.

In order to understand how fire grows and spreads, it is important to remember that heat can be transferred by conduction, convection and radiation and each will be involved to a greater or lesser extent depending on how well the fire is developed. These transfer processes are

important in both the growth of the fire in each space and the generation of the ignition source for the spread to other parts of the building.

It is also important to know that the objects and materials that combust give off not only smoke but also flammable volatiles such as gases which, in the presence of oxygen, burn and we see flames. This adds greatly to the heat evolution.

Fires normally start small. Only objects in contact with the original ignition source start to burn. Gases, heat and smoke will be given off. These will rise in convection currents, heating other possible fuel sources which start to give off gases, and these all collect under the ceiling where the surface area of this flammable material becomes much greater. If there is still oxygen to support the fire, when the flames reach the ceiling and the temperature is great enough the gases will ignite. The resulting increase in radiant heat will now cause ignition in most of the remaining flammable objects and materials (particularly surfaces, or linings) in a very short space of time. This point occurs at around 600°C and is known as flashover; the fire is then fully developed. If there is insufficient oxygen the gases will not ignite, but the objects that are on fire will continue to smoulder and give off heat. The introduction of oxygen as the door is opened will cause the sudden ignition of the gases, causing the phenomenon known as back draught.

The temperature in the space now rises rapidly and the potential for the fire to spread becomes much greater. Containing the fire at this point depends on the fire resistance of the elements such as walls, floors and the building structure.

By following the development of the fire as described in this way it is possible to identify the technical solutions that might be used to make sure people escape and that the integrity of the building is maintained throughout the fire.

Fire prevention

It is clear that the best defence against fire is to prevent ignition. Identifying the possible sources of ignition and isolating them from a significant supply of fuel will accomplish this. If the activities or processes undertaken in the building produce heat then they should be carefully designed so as to be isolated some distance from any combustible material. These can often be easily identified as heat-producing appliances and flues. Minimum distance of timber components from flues is a good example of fire prevention. However, another major threat is the malfunctioning of any energy distribution service such as electricity and gas. Malfunctioning electrical systems are a common cause of fires. If the electrical current arcs it provides a direct source of ignition at high temperature, causing immediate ignition of surrounding fittings or construction. In extreme risk areas, low-voltage electrical supply may be specified to minimise the risk of electrical fires.

Ignition from accidental sources such as discarded cigarettes is difficult to control by strategies involving the choice and detailing of the construction. However, treating fabrics with fire retardants has had major success in limiting the start of fires from such accidental sources.

Fire detection and escape

If a fire does start then there should be a maximum chance of it being dealt with in the initial phase of the time–temperature curve. This involves early detection and fast extinguishing. These two functions are most often associated with the services provided in a building. Both detection and alarm can be either manual or automatic. The simplest means of detection and alarm that has been used is the provision of a bell (or triangle) to be rung by hand by the person who detects the fire. Alarm may still rely on an individual activating a building-wide alarm system automatically by breaking a glass disc in front of the alarm switch. Increasingly an automatic detection system will be specified using heat and smoke detectors to set off the alarms.

Having detected the fire and raised the alarm two activities have to be considered together: the evacuation of the people and the provision for fire fighting.

Given that the fire has been detected during the growth phase indicated on the time–temperature curve, heat will still be relatively low, although smoke may build up very quickly. If the fuel is chemically complex the smoke may be acrid or even toxic. Escape has to be a priority. The majority of deaths in fires are from asphyxiation by hot smoke rather than burns due to the heat.

However, with very early detection it is possible to smother a small fire. The traditional manually applied material would be sand or a fire-blanket, but now more probably with the contents of an extinguisher. There is a link here with the operational management of the building. The contents of all extinguishers are not the same. Products of water, foam, powders and CO_2 are available for different types of fire; clearly water should not be used on electrical fires or even fat fires. If initial fire fighting equipment is to be provided then personnel have to be trained in its use and the equipment checked regularly.

Automatic initial extinguishing can be achieved with sprinkler systems (or their gas equivalents as long as evacuation can be achieved first). These include a detection system that automatically releases the extinguishing medium, water in the case of sprinklers. In sprinklers each sprinkler head is its own detector (heat expands a liquid in a glass stopper which breaks and allows water to escape from that head only), so extinguishing can be as local or widespread as the fire. Even if the system does not extinguish the fire it gains valuable time in that it considerably reduces the production of both heat and smoke, prolonging the initial phase on the time–temperature curve and the opportunities for escape. It does, however, require a large volume of water if the control is expected to be widespread as in large open warehouses. Here the public water supply may not be able to deliver at the required rate. Large storage tanks with pumps may have to be provided to maintain flow should the sprinklers ever have to be used.

As soon as detection and alarm have been established evacuation should commence. Again the link between construction provision and occupation management is very clear. The design of the building must provide the means of escape, while it is the responsibility of the occupation management to ensure that the routes are known and maintained unobstructed.

The design of the building for means of escape is initially a case of planning routes and determining the size of these to ensure a speedy and panic-free evacuation. Everybody should perceive that they have an equal chance of escape. Initially anybody in the space in which the fire starts has to escape. They have to be able to move to areas that are on the escape routes directly linked to the evacuation point. Here they may meet people from an as yet unaffected part of the building also heading for the evacuation point. While most of these escape routes will utilise the everyday corridors and stairs, additional provision is often necessary due to the fact that everybody would be using them at the same time. It is also necessary to provide alternative directions of escape so individuals do not have to go through the fire to reach evacuation points. The distances and size of these routes have been determined empirically and are now provided in codes and regulations, but computerised modelling can also provide predictions of behaviour in these early stages of evacuation.

These escape routes need protection. They need protection from heat and smoke if escape is to be effected. This protection will be provided primarily in the choice of materials and detailing of the construction that encases the route and ultimately in the stability of the structure of the building itself. In order to think about the properties of the material chosen and the appropriate detailing it is necessary to see how a fire spreads from its initial ignition to other parts of the building.

Building fabric fire performance

Fire development and fire spread are the two areas where the fabric of the building makes most of its contribution to fire design. The floor, wall and ceiling coverings, or linings, influence the development of the fire, and then fire resistance of the elements, both structural and non-structural, influence the spread. The role of the fabric can be identified as significant in the following two broad areas:

(1) Lining materials
(2) Fire-resistant elements
    o Internal wall and floors
    o Structural members
    o External walls and roofs

While linings are significant in the initial stages of the growth of a fire, within each space the fire-resistant elements are important as fire barriers and maintaining structural stability. Fire barriers also slow the spread of smoke within the building and, in some cases such as compartment walls and floors and in protected shafts, contain the fire within sections of the building. In the design of party walls and external facades the aim is to ensure that the fire does not spread to adjacent property. Structural integrity needs to be maintained throughout the fire where the question of its potential performance in any rebuilding might be considered.

Fire performance of linings

The choice of linings can affect the time available for escape by slowing down the growth of the fire, but they may also contribute toxic gases and/or flammable volatiles which might add to flashover gases or create hot droplets that might be a source of ignition. The properties of the materials for fire performance for the linings to the floors, walls and ceilings can now be identified. Combustibility, surface spread of flame and fire propagation all indicate how quickly, if at all, the lining will assist the development of the fire. Smoke and droplet (burning particles) tests along with measurements of contribution to flashover gases give a complete picture to evaluate the choice of lining against fire performance criteria.

Fire resistance of elements

Having chosen materials for the linings the next concern is the spread of smoke to escape routes and the fire igniting material outside the room. Controlling spread is the role of the fire-resistant elements. Performance is not just dependent on materials properties but the design of the whole element, be it wall, floor or structural components. This is recognised in that while we talk about lining materials we also talk about the fire resistance of elements. Whole assemblies have to be tested including all the detailing, joints and fixings (and loading for structural elements) for a performance rating for fire resistance.

Fire-resistant elements may have the role of a barrier to smoke and fire and/or be a part of the building structure, the collapse of which would not only allow the spread of fire but also cause physical danger to occupants escaping and members of the rescue services.

There are three ways in which fire-resistant elements might fail to ensure a fire is contained:

- Stability (of structure)
- Integrity (of barriers)
- Insulation (of barriers)

The 'stability' of a structural element may be affected if the fire reduces its load carrying capacity to the point where the element collapses under the load it has to carry at that stage in the fire. The 'integrity' of a barrier may be lost if the element cracks, shatters or just burns away. This includes jointing materials that may burn or melt, or fixings that fail under the influence of the fire. The 'insulation' of a barrier must be sufficient to keep the other side of the element from reaching temperatures that will cause ignition. More correctly, 'insulation' should be seen as heat diffusion. High thermal mass or endothermic reactions (chemical changes that absorb heat from the fire) will have a similar effect of maintaining the other side at temperatures below ignition for the duration of the fire even though in thermal resistance terms the material of the element may be considered a poor insulator.

Internal walls and floors can contain fire and smoke to control the spread of a fire. Not all of these barriers need the same degree of resistance to the passage of fire even within the same building. The performance requirements will depend on two factors. First, the nature of the occupancy including the total numbers of people in each area and their mobility at the time at which they have to use the escape routes. This will determine the actual time it will take to evacuate everybody to safety. Second is an estimate of the speed of growth and ultimate ferocity of the fire as it follows the pattern illustrated in the time–temperature curve. This will determine safe evacuation periods.

Another consideration that will determine the fire resistance of elements is the size and configuration of the building. In order to limit the fire resistance requirements for individual elements it is possible to design the whole building as a number of fire compartments separated by compartment walls and floors designed to contain the fire completely. While the specification of these elements may be particularly onerous they do reduce the demands on the other elements inside the compartment as the risk to both people and property is significantly reduced.

Fire may spread more easily at openings in the floors and walls acting as fire-resistant elements if not correctly detailed. Openings are potential breaches in the integrity or insulation of the element. Two types of opening can be identified: doors (and opening windows) and spaces for services to pass from one area to another.

Doors need particular attention, as they have to contain both heat and smoke from escape routes. Not only must the door itself have properties of integrity and insulation but also the gap around the door is particularly vulnerable in spreading the fire to the outer surface of the door. The dimension of the closing edges and the coverage offered by the rebate become a particular aspect of detailing. Closing edges on double doors may have to be fitted with intumescent strips. These materials expand on heating, sealing gaps and ensuring integrity and offering an insulating layer against heat transfer.

The small spaces created around the services passing through walls and floors need to be filled or 'stopped' with a fire-resistant material to maintain the integrity of the element.

The services are in fact often carried in voids either above the ceiling or below the floor. The convenience of these voids is that the services may pass unhindered above or below the rooms independent of the partitions between rooms. In fire design these are known as cavities that have the potential to spread smoke and fire rapidly to other parts of the building. There is a need to provide barriers in these cavities to prevent the spread of both smoke and fire. These cavities should not extend across compartment walls or through compartment floors.

Another danger is spread via vertical shafts. These shafts are provided in buildings for circulation (stairs and lifts) and for services. Once breached by fire they will act as flues and transmit heat and smoke to higher floors very quickly. The fire resistance of these shafts becomes an important aspect of the fire design. Protected shafts between compartments have to have walls with high resistance ratings and the openings such as doors and service entry points have to be fully protected. Since the stairs are probably escape routes, preventing the intrusion of smoke may be a major performance requirement possibly by the introduction of positive pressure.

So effective are vertical shafts and cavities that even the cavities in cavity wall construction provide a pathway for the rapid spread of fire to higher floors. This leads to the need to provide barriers in these cavities if the internal skin has limited fire resistance.

As well as establishing barriers to control the fire it is necessary to ensure that the structure of a building does not collapse or damage the barriers for the duration of the fire. There may also be some consideration of the integrity of the structure at the end of the fire for its use in any rebuilding work.

Structural materials in fire

Elevated temperatures change the properties of many materials, including strength, so stability of load-carrying elements may be lost without any apparent physical fire damage to the element. It is necessary to understand that structural sections exposed to fire may not burn, but the rise in temperature may cause expansion and/or a significant loss of strength. The section may deflect and even collapse, as it is no longer strong enough to carry the loads.

The mechanisms are different for the major structural materials. Steel has a high thermal conductivity and low thermal capacity which leads to a rapid rise in the temperature when exposed to fire. There is, therefore, a rapid loss of strength as steel has lost something like 50% of its yield strength at 600°C, well within the temperatures reached in a building fire. Steel can be fire engineered but is usually protected from the heat. Casing or spraying the steel with either an insulating material or a material that can absorb the heat can achieve this. Absorbing can be achieved by either high thermal capacity or endothermic reactions where chemical changes in the materials use the heat of the fire. These casings and spray-applied materials must remain attached and not crack during the fire. Other strategies that have been used include filling the structure with water to absorb the heat or putting the structure on the outside of the building, but these options have not gained widespread application.

Reinforced concrete has a high thermal capacity so, unless the sections are thin, the rise in heat during a fire is slow other than near the surface. If the surface cracks or spalls or if the steel reinforcement that is near the surface is allowed to gain heat, the section may fail. This focuses attention on the concrete providing cover to the reinforcement. The aggregates in the concrete and the depth of cover are the specification items of significance to the fire resistance of reinforced concrete.

Timber has a completely different mechanism associated with achieving fire performance. Timber burns, but the charred products of combustion remain attached to the section and 'insulate' the unburnt timber from oxygen and heat. Burning slows considerably as the fire progresses. If unprotected structural timber is specified oversize to account for a charred layer the timber will carry the load for the duration of the fire but will need to be replaced in any rebuilding.

Fire performance of external elements

Having specified internal barriers and the structure so that they perform during a fire there remains one other major threat. If the fire breaches the external envelope not only will the fire gain a source of oxygen but also the fire may spread between adjacent properties. There are two ways that this spread between properties might happen. The first is spread along external surfaces to adjacent buildings, particularly along roof surfaces. The second is directly from areas of the façade breached by the fire. While spread along external surfaces implies that the buildings are connected, breaches in the façade imply a distance to the adjacent building. If this space between the buildings is large enough the risk of spread is eliminated altogether. Up to this point, at least some of the façade must act as a fire-resistant element. The closer the buildings are together then the greater the area that must be fire resistant. The codes and regulations give guidance on how much unprotected area can be allowed against notional boundaries between buildings.

Specification of fire resistance

All these performance requirements need to be specified in terms of time. There remains the problem that while evacuation can be established in real time, the actual time an element will hold back the fire depends on the ferocity of the fire. For comparable performance between elements, testing must use a standard fire, one with a predetermined time–temperature curve. Hence the times quoted for fire resistance are not real time but standard test times. The required 'fire resistance' has to be chosen taking into account both the required evacuation real time and the ferocity of the fire defined by its time–temperature curve.

Fire-fighting provision

There remains one other area of consideration for the design of the building; that of the provision for fire fighting by the professional fire service. Earlier discussion covered the provision of equipment for fighting the fire by the occupants early in its development. Once a fire starts to develop this equipment has limited value. It becomes necessary to give access and provide services such as hose reels, wet and dry risers and foam inlets for the use of the fire brigade. This will require consideration to be given to the access of fire appliances into the grounds of the building and, possibly, on large buildings, even specialised personnel access in the provision, for example, of fire lifts. Like so much fire provision this is a matter for initial design, but it also has to be managed 24 hours a day for the life of the building.

Threats from within society

Some threats come from the society within which the building is designed to operate. Like the natural threats (acts of God) and fire, they are threats to both person and property. In most politically stable societies direct threats to the person from strangers are relatively low unless associated with a threat to property. These threats are most usually associated with crime, possibly including the extreme of terrorism. They are characterised by the means individuals are willing to use to act out the threat. These are complex sociological problems and it has been suggested that the quality of the built environment can play its part in reducing the threat as well as providing defence against it.

The nature of the threat is highly dependent on the activity to be housed in the building. For the most part it is concerned with the intrusion of an external threat, but there are situations where the threat is being housed, e.g. prisons, and the risk is to the activities being carried on outside the building.

Whichever of these is the case, the approach to design is likely to follow that identified for fire: deterrent or prevention, followed by detection and alarm, followed by the provision of resistance to the forces of the threat in order to limit injury to persons and damage to property. Unlike fire, however, the perpetrator of the threat has reasoning and therefore defining any set pattern of attack is difficult. For instance, there is for each perpetrator a fear of detection and an assessment of the risk of being caught.

Assessing the threat involves assessing determination and the means the perpetrator is willing to use. Petty crime against households may be deterred by the provision of good locks and simple common-sense occupation management to stop opportunistic crime. The determination to burgle any specific property is low and the means that will be used will not involve large forces. At the other end of the scale, terrorism is characterised by having specific targets, high levels of determination and access to considerable force to inflict the damage necessary to breach the defence offered by the fabric and structure of the building. An interesting third example would be a 'cat' burglar who may target specific objects, plan and execute the intrusion with great determination but operate with very little force. Devising strategies and the performance of the construction solutions chosen is dependent upon understanding the mind and means of the perpetrator.

Personal accident

The main focus of the analysis of a safe environment has been possible external threats but there is also the need to reduce the risk of personal accident. This determines much of the dimensioning and detailing of ramps, staircases, balustrades and handrails. Another example is specifying floor finishes that are not likely to become slippery when wet.

The private environment

The origins of the requirements for privacy are the conventions of society. Unlike safety, the needs do not arise from the existence of a threat but from a sense of what is right and wrong, what is acceptable or unacceptable social behaviour and organisation. To understand the needs of privacy it is necessary to understand the ways of the society itself and the beliefs of its people.

The concept of the family dictates much about the form of dwellings. Management style and organisational structures will determine much about the form of commercial buildings. This is linked to the concern for image and status, which was considered as an aspect of appearance but is associated with other aspects of the performance of the division between spaces to carry out separate activities.

The feeling of privacy comes not from the absence of the sense of impending physical danger but from the close proximity of others invading 'personal space', even if no physical barriers exist in a room. However, social convention often requires physical separation, involving limiting the attention of others seeing or hearing or in any way being aware of the activity associated with the use of that space.

As such, the major requirements for privacy are acoustic and/or visual separation. The need for privacy adds to the performance requirements of the dividing elements of the building. Although predominantly walls, it includes the construction of floors, ceilings and service distribution routes where 'cross-talk' can be a problem.

The analysis and evaluation of the performance of visual separation needs no complex theory or testing other than that of direct observation. Predicting acoustic performance is less direct and has been considered earlier in this chapter.

Summary

(1) The aspects of the environment that need to be considered have been presented as dryness, warmth, light, acoustics, cleanliness, safety and privacy. The analysis of these seven areas is the focus of this chapter.
(2) The building is a system of active services and passive fabric whose role and contribution (function and performance) is decided in the design.
(3) For each of the aspects to be considered it is necessary to have a dynamic description of both the external conditions and the response of the construction to those conditions.
(4) Each aspect of the environment that might lead to failure in the proposed construction must be analysed to ensure the performance of the whole.

11 Physical Behaviour Under Load

This chapter recognises the fact that, while all parts of construction are subject to loads, even if it is only their own self-weight, there is a specific need for a building to have a system of structural members jointed together to transmit the loads to the foundations. It is necessary to determine the way the loads are applied and how the support is provided for each structural member together with the way they are jointed to one another. This will establish the pattern of internal forces that will generate the stresses in the materials chosen. It is suggested that visualising the distorted shape of the structural members will indicate the pattern of internal forces. The system must contain sufficient members and be jointed in such a way as to remain stable, particularly against horizontal forces such as wind loads. With knowledge of the stress/strain relationship of the material this pattern of internal forces gives the clues to the efficient shape of the structural members and the economic size.

Forces, external and internal

As soon as any part of a building is created it will be subject to forces even if only from its own weight, and therefore has to have some strength and adequate fixing. There are some elements of the building, however, whose function would be recognised as part of its structure. They maintain the shape of the building and transfer loads from one part to the next to reach an existing stable support, normally the ground, which has to take the accumulated loads and therefore becomes part of the structural system.

These external forces, or loads, are predominantly generated by the weight of the construction and the objects and people inside the building. However, there are also environmental loads such as wind and snow which create forces on parts of the building exposed to the elements that have to be transferred successfully to the ground.

These external loads create internal forces within the structural elements. The nature of these forces and the mechanism by which resistance can be provided become the focus for the analysis of the physical behaviour under load. These elements of structure have to ensure the building remains stable, has limited distortion and does not break. Instability, distortion and

breaking are the possible failures that will have to be identified and performance levels established against which an evaluation could be made.

The initial design task is to identify a system of structural members and supports that, when jointed together, provide a stable structure against the loads that will be applied to them. Loads applied to one member in the system have to be transferred to a support through some form of joint. The load being transferred through the joint effectively places a load on the support that becomes the next member in the system. This load path has to be traced through the structure. Each member and joint must be capable of resisting the accumulation of loads being transferred through the structural system.

It is important at this initial stage to ensure that the structural system has all the members necessary to ensure its overall stability. The initial effect of the loads will be to displace or rotate the members to which they are applied. This can be resisted at the joints between structural members, but often no such resistance exists. It is not until the additional stability members have been identified that the overall structural system has been devised.

Once the structure has sufficient joints and members to achieve overall stability then the loads will appear to distort the structure. This distortion arises because the externally applied loads induce forces inside the members and joints that stress the materials of which the structure is made. When materials are stressed they undergo dimensional changes known as strains that cause the members to distort. As stress increases in the material it will ultimately cause the member to break.

Understanding structural behaviour and undertaking the design of structures is concerned with establishing stable structures and the analysis of the internal forces and the stresses the loads induce, along with the strains that arise causing the distortions.

The outline above follows the general approach to the analysis of physical behaviour established in Chapter 9:

(1) The loads define the conditions under which structures have to perform.
(2) Failure or loss of performance can occur if the loads
  o disturb the structure, making it unstable
  o distort the structure too far so it cracks finishes, transfers loads to non-loadbearing members or looks and/or feels unsafe
  o destroy part of the structure by breaking members or connections at the joints.
(3) The proposal or suggestion is for an arrangement of structural members with choice of materials and approximate sizes to support the loads.
(4) An evaluation of the response of the proposed members to the loads is by analysis of the distortions and the distribution and size of the forces within the structural members to see if they will maintain the loads without failure.

It is now necessary to look at each of these stages in more detail.

Loads and loading patterns

Loads on the building come from the weight of people and the objects in the building, the weight of the building itself and from the environmental conditions to which the building will be subjected. What they have in common is that they all apply forces to the construction, which have to be resisted if the building is to continue to remain stable and then not distort excessively or even break.

The most usual way to classify loads is as follows:

● Dead loads – the weight of the construction itself
● Imposed loads – from the activity within the building and environmental loadings from snow
● Wind loads – an environmental loading from the wind

Before the characteristics of each of these types of loads can be appreciated it is necessary to identify what needs to be known about a load before it can be evaluated as a force on the building.

The following represents a definition of a load:

● Magnitude or size
● Distribution
● Direction

Magnitude is the total weight or force to be taken by the structure, while distribution identifies the pattern on the member. Distributions are often seen as a single point, uniformly distributed or possibly moving and impact loads. Direction completes the specification. Most loads on a building act vertically downwards under the influence of gravity, some will act at other angles (thrust at the bearing of an arch), with some, such as the wind, even acting vertically upwards under certain conditions.

It is now possible to see the basis of the classification. The size of dead loads can be calculated with some certainty as can the pattern of the loading and its direction. Imposed loads can be identified with less certainty. While magnitude can be established within certain limits, the patterns of loading need to be established to identify the range of design loading conditions. There may be some objects to be used in the building such as machinery or storage arrangements that are so much larger than any of the other components of the imposed load that they would be treated separately, with their position on the floor defined with the characteristics of a dead load. However, the majority of imposed loads can be brought together as static uniformly distributed load defined by the notional use of the building, e.g. domestic or office.

Wind loading is less easy to define. Its inherently dynamic nature coupled with its variability in intensity and direction make a definition of size, distribution and even direction problematical. As wind speeds up as it flows over and around building it creates low-pressure areas, resulting in uplift forces particularly across roofs. If vortices are created particularly at edges, even greater suction forces can occur. In order to provide a definition of the wind load (and hence the forces on the construction) that can be used in a structural analysis, certain simplifications have to be made to obtain an equivalent static load. These simplifications must, however, ensure that the effect predicted is at least as demanding as the effect that will be experienced by the real structure under the complexity of the real environment. Procedures for this are now well established and published in codes and standards.

Another condition that has to be allowed for is the possibility of an accident causing a local structural failure that leads to a more general failure of a major part of the building. This is known as a disproportionate collapse in that the final proportion of the building affected is disproportionate to the local nature of the original failure.

Some buildings, however, will have to be considered for special loading, such as impact from aircraft for high-rise buildings and blast for security buildings or those deemed to be at high risk from terrorist attack. This serves to highlight the importance in performance specification of defining the conditions that should be designed for, either as the limits considered as normal or in terms of extreme yet potentially disastrous conditions. However, the majority of buildings are only designed for dead, imposed and wind loads plus measures against disproportionate collapse. Designing for extreme loading in most buildings would not be seen to have value in the inevitable extra building cost, as the probability of them occurring would be very low.

Basic structural members

A limited number of types of structural members can be employed in buildings. They act either to span around spaces or to provide support or stability to hold the spanning elements in place to generate the spaces within the building.

While the common names for these structural elements – walls, columns, beams and slabs – give some clues to the position in the building and their basic form, understanding their structural behaviour needs closer analysis:

- Where the loads are applied
- What support is provided (to make stable and stop disturbance)
- How the stable member distorts

By visualising the distorted shape it is possible to identify the distribution of internal forces that would break the member.

Figure 11.1 illustrates perhaps the most straightforward structural members: the strut and the tie. The strut (most commonly identified as a column) is loaded and supported in such a way that it experiences compression, identified from visualising the shortening caused by the loads. By contrast, the tie is loaded and supported so it goes into tension, indicated by visualising lengthening.

Figure 11.2 illustrates the beam, a spanning member where the loading and support pattern causes the member to bend. Two beams are shown in Figure 11.2 each with their distorted shape, indicating that where the material in the beam shortens it experiences compression and where it is stretched it experiences tension. The maximum lengthening and shortening will occur in the top and bottom of the beam, with no distortion being experienced along the neutral axis. All bending members have this pair of tension and compression forces acting together across the depth of the section. All bending members have a second pair of tension and compression forces known as shear. This time the

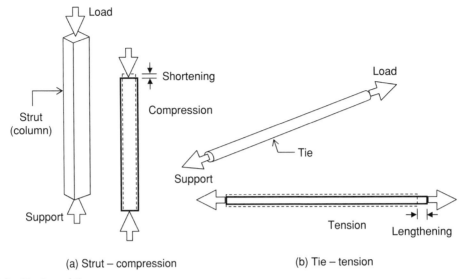

Figure 11.1   Struts and ties.

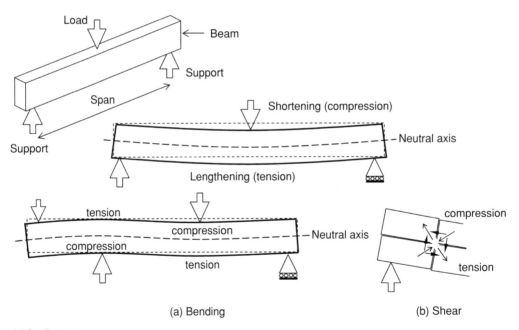

Figure 11.2   Beams.

forces are greatest at the neutral axis of the beam where they act diagonally across the material of the member.

Struts, ties and beams have direct compression, tension and bending forces respectively.

The other two major structural members, walls and slabs, can have different load and support patterns giving different sets of internal forces. Figure 11.3 shows how a wall can be in compression when taking loads through the building,

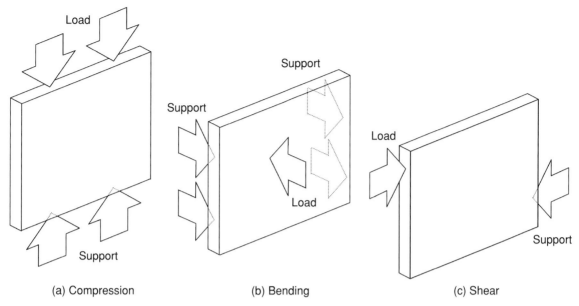

Figure 11.3 Walls.

but may be in bending if resisting horizontal forces such as wind on the façade or ground pressure as part of a basement structure. The shear-loading pattern is created when a wall is used to stabilise the racking effect of wind forces on building structures, as discussed later in this chapter.

Similarly, slabs can be subject to different internal forces depending on the loading and support arrangements. Figure 11.4 shows that this is mainly a distinction between suspended and ground-supported slabs. Suspended slabs will be in bending. The illustration shows support on only two edges, creating a one-way span, but it is possible to support the slab on all four edges, creating two-way spanning, to reduce sections or offer potentially greater spans. Ground-supported slabs carrying uniformly distributed loads will experience compression from the applied loads. When slabs are placed under columns and walls as foundation, the forces they experience depend on the ratio of the depth to the width of the foundation. If the foundations are comparatively wide, the foundation must act in bending for the load from the column or wall to be transferred evenly to the ground. When the slab becomes narrow the predominant force experienced by the foundation is a punching shear.

Lateral distortions

Figures 11.1 and 11.2 show the shortening of the strut, lengthening of the tie and deflection of the beam as distortions that arise as a direct result of the application of the load and in the direction of the load. These are not the only distortions that occur. Lateral distortions may be experienced in slender members where compressive forces are experienced. These are known as lateral distortions because the member bends in a direction at right angles to the direction of load. Slenderness, as the name suggests, is primarily a function of the height (or length) to width ratio, but it is also dependent on the way the member is connected with the building structure. The effect of the connection will be explained later.

Observation of the internal forces that develop in the members discussed in Figures 11.1 to 11.4 shows that compression zones develop in all types of members except ties. However, not all of these members become slender in a direction at right angles to the load. Again, by observation, perhaps the most obvious potentially slender

Figure 11.4 Slabs.

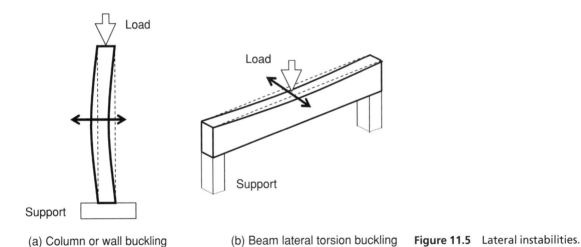

(a) Column or wall buckling   (b) Beam lateral torsion buckling   **Figure 11.5** Lateral instabilities.

compression member is the strut and the potentially slender compressive zones that arise in beams. The lateral distortions, known as buckling, which might occur in struts (column or wall section) and beams, are shown in Figure 11.5. In columns the buckling induces bending as the whole section is in compression. However, in beams the compression zone buckles while the tensile zone remains straight. This induces a twisting action known as torsion. The whole lateral distortion effect is known as lateral torsion buckling.

Perhaps less obvious is the possible buckling associated with shear. In thin web sections in beams the principal shear compression force shown in Figure 11.2 may develop buckling. In thin shear walls, shown in Figure 11.3, the buckle will occur across the diagonal, giving a potential bending failure similar to that of the buckling strut.

It is important to realise that the buckling distortion does not replace the distortions of the direct loading, but when it occurs, the structural member experiences both. The two sets of internal forces develop together. For instance, in a column already experiencing compressive forces the buckling adds bending, increasing compression on one side but reducing the compression where tensile bending forces from the buckling develop.

Eccentric loads

In identifying all the distorted shapes and internal forces so far it has been assumed that the loads have been applied and support has been provided at the centre line (on the axis) of the member. These are known as axial loads acting through the centre of struts and ties and along the centre lines of beams and walls. Although it is desirable for the economy and efficiency of structural members to be axially loaded, it is not always possible. Loads that act off the axis are known as eccentric loads, eccentricity being a measure of how far the load acts away from the axis.

The application of eccentric loads may cause the member to overturn if the dimensions and/or connections are not sufficient for it to remain stable. However, assuming the member is not disturbed in this way, eccentricity creates a further distorted shape and therefore changes the distribution of internal forces. In columns and walls they induce bending. In beams if the ends are restrained from rotation the effect is torsion, similar to lateral torsion buckling.

Like lateral distortions the additional distortions caused by eccentricity do not replace the direct loading effects but add to them. The effect in the column is, therefore, as described for buckling. Already in compression, the bending forces will increase compression on one side and the bending tensile forces will relieve some of the compressive force on the other. Because eccentricity induces a bending effect this is likely to be the direction in which buckling occurs, given that the member has the same slenderness in all directions.

Curving (and folding) structural members

The members discussed so far have all been straight or flat. Curving the member may increase the manufacturing and assembly cost but does provide structural advantages and may therefore be worth considering. The structural advantage is associated with spanning members. To develop sufficient resistance to span in bending, the material in compression and tension from the direct bending effect (i.e. not the shear) has to have sufficient separation to develop the moment of resistance. This makes beams and slabs deep if the spans become too large. This increases the depth of the structural members and, unless some of the solid section can be removed, increases the self-weight, adding to the dead load to be carried by the beam or slab.

Advantages can be gained from curving both in the direction of the span and at right angles to the span, but for different reasons. Consider the members in Figure 11.6. These members that curve across the span are known as arches and suspension cables. Two significant aspects should be noted. First, they require both a horizontal as well as a vertical support and, second, the internal force becomes pure compression and tension respectively: they are effectively curved struts and ties. This allows materials and/or components naturally strong in either compression or tension, but not both, to be used in the construction. Less material is needed and more material options are available. When slabs are curved like arches they are known as shells and where allowed to hang suspended they are known as nets or fabric structures (tents).

Curving and folding at right angles to the span gains a different advantage. It gives the depth that is needed to act in bending from thin sections. The most common use of this technique is for roof sheets or decking, as shown in Figure 11.7. Thin sheet materials can span many times that of the flat sheet form. Again, less material is used, but there is a manufacturing cost associated with the need to form and control the folding or curving. Curving in this direction is still a

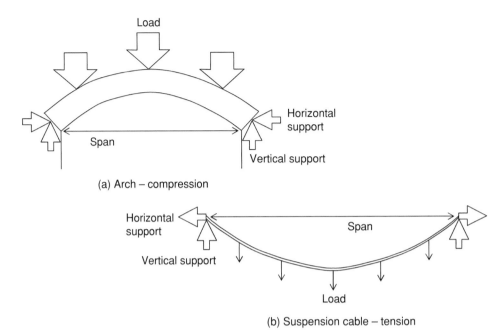

(a) Arch – compression

(b) Suspension cable – tension

Figure 11.6 Curved structures across the span.

Figure 11.7 Curved (folded) structures in the direction of the span.

bending action and therefore the supports are only required to take vertical forces.

Structural connections

In the analysis of the five basic members shown in Figures 11.1 to 11.4 there is an implicit assumption about the way the member is connected to the support. All the distorted shapes assume that the member can rotate if it bends and can shorten or lengthen freely in response to the distorting effect of the loads. Whether this can happen or not depends on how the connections are detailed.

There are advantages to restricting the movements at the connections, particularly that of rotation, to induce different distorted shapes to redistribute the internal forces to reduce the size of the member. This does sometimes increase the manufacturing and assembly effort and therefore has a cost to be set against the saving in section size.

In order to think about the effect of restricting movement at the connection on the potential stability and distorted shape of the member, three 'pure' actions can be identified and are shown in Figure 11.8. They are shown diagrammatically as they are idealised structural joints. The detailing to gain pure action is often difficult and expensive. The roller joint allows both rotation and freedom to move against the shortening (or lengthening) effect. Pin joints allow rotation and rigid (fixed or moment) joints allow neither.

Figure 11.8 Structural action of connections.

(a) Roller   (b) Pin   (c) Rigid (moment)

(a) Rigid connection between beam and column   (b) Column with rigid connections

Figure 11.9 Effect of rigid connections.

Figure 11.9 shows the effect of making the joints rigid on two of the conditions already considered. The first is making the joints rigid at the end of the beam. The support now has a rotation (or moment) action to resist. If the support were a column with no balancing force from the other side (for instance from a similar beam with similar loading and similarly fixed), the column would go into bending. However, the internal forces in the beam are redistributed along the beam and the maximum force in the beam is reduced, but bending forces appear at the supports where there were none when the joint was assumed to be a pin. Tensile forces appear at the top of the beam at the support; the balancing compressive force being at the bottom. This redistribution of the bending along the length of the beam reducing its size is the basis of the portal frame, which can achieve long spans

economically where a simply supported beam could not.

The second example in Figure 11.9 is the effect on buckling in a column where the connections are fixed. Because of the moment resistance experienced at the end of the column, the effective length of the column is reduced and hence the slenderness is reduced. This column will take greater loads before it buckles.

In reality, most connections are detailed for ease of manufacture and assembly rather than to ensure a pure structural behaviour. They give these freedoms of movement to a greater or lesser extent and therefore the real instabilities and distortions may not be exactly those assumed and different internal forces will develop. However, most simple connections behave near enough to a pure form (normally pinned) for the assumption to be safe to use with the practical detailing.

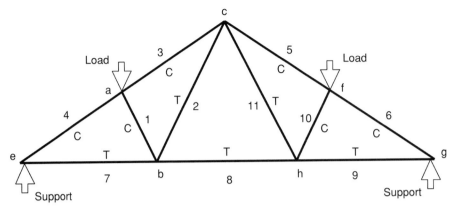

Figure 11.10 Pin-jointed frames – the truss.

If there is doubt about the degree of fixity (often known as semi-rigid joints), an estimate of the effect has to be established so that a more realistic distribution of internal forces in the members can be calculated. If the structural advantage of making the pure behaviour outweighs the additional cost of manufacture and assembly then details have to be devised to ensure the pure behaviour is achieved.

Grid members – pin-jointed frames

All the structural members considered so far, flat or curved, have been solid. Curving across the span was worth considering for bending members to generate pure forces (compression or tension) and reduce the dead loading that occurs as beams and slabs are designed to span greater distances. Pin-jointed frames are another way to achieve the same ends. This is a very versatile idea and many forms of these frames are used, both as plane frames (having little width like beams) and acting in three dimensions like slabs, both flat (double-layer grids) and curved (single-layer grids as domes) structures.

The truss is one of the simplest forms of the pin-jointed frame and will serve to demonstrate the principles. It is shown in Figure 11.10. This is a spanning member with loads and supports shown and all pinned joints are assumed. Taking each member of the truss one by one it is possible to trace the load and support of each to

determine its distorted shape. The load applied to node a is directly transferred into member 1 which gains its support from node b. This will shorten member 1, putting it into compression. It has to be designed as a strut (note the load will be axial but it may buckle). At node b, forces are transferred into member 2. Consider the effect on member 2. A downward force is experienced which is supported at node c. Member 2 is stretched: it is in tension and therefore acts as a tie (note not subject to buckling). Forces at node c now pull down on members 3 and 5, pulling them together rather like the action of an arch. This puts compressive forces into these members that are continued through members 4 and 6 to nodes e and g respectively. This generates a need to support a horizontal force at e and g and members 7, 8 and 9 give mutual support to e and g. These all become ties. By the same logic, member 10 is a strut and 11 a tie. All the forces are resolved within the frame as either struts or ties. The supports now only experience vertical loads, assuming the connection of the truss to the support to be a pin.

This analysis is based on the assumption of the truss acting as a pin-jointed frame. There is, however, a need for the connections to be practical while being able to assume a structural pin joints action. Figure 11.11 shows a typical detail in steel and in timber of the connections at node a. The illustration shows that in reality members 3 and 4 are all one piece, but in this case the relatively low stiffness of the members

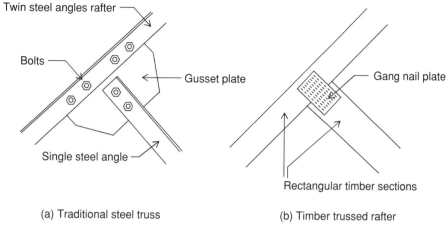

(a) Traditional steel truss   (b) Timber trussed rafter

Figure 11.11   Detail of joints in pin-jointed truss.

themselves means that the assumption of pin jointing is safe.

Building structures – wind stability

When considering structural members, so far the loads have all been vertical, essentially from the dead and imposed loads on the building. Actions of forces other than vertical have been identified, but these have come when members transfer their loads to other members as the loads find their way to the ground. When visualising a whole building structure capable of carrying these vertical loads, it may not be stable against wind loads. Only when the whole building structure is identified can the effect of wind loads be considered.

Wind loads can act vertically. One of the most destructive aspects of wind to the fabric of the building is the negative, uplift forces on roof structures as the wind speeds up as it is forced to go round and over the building. Where roofs are over large open spaces the wind can get into the building and create positive pressures under the roof, increasing the effect of the uplift forces. However, the major initial concern to the overall structure safety is the horizontal forces acting on the sides of the building. These can be both positive and negative forces, often being positive on one side and negative on the other, having

a cumulative effect on the structure. One further aspect of the wind is that, while it may have a prevailing direction it will sometimes blow from other directions and therefore the building must be designed to withstand wind from all directions.

Wind or overall stability of a building is a major design criterion. Additional members may be necessary to stop wind disturbing the structure. These members and/or connections allow the forces to be distributed through the structure and safely transmitted to the ground.

The major structural systems used for buildings are loadbearing walls and framed structures which also include most of the long span roof solutions. The predominant instabilities and the measures to overcome them vary between these two structural systems. However, there are only a limited number of strategies to overcome these overall instabilities, which can be used alone or in combinations:

- Increasing dimensions (against overturning)
- Changing joint type (from pinned to fixed)
- Providing additional support members (buttressing, bracing, etc.)
- Changing shape (curving, folding, etc.)

Some of the alternatives used for walls and frames are shown in Figure 11.12. The major instability failure for walls is overturning. Widening and embedding are options most likely to be used

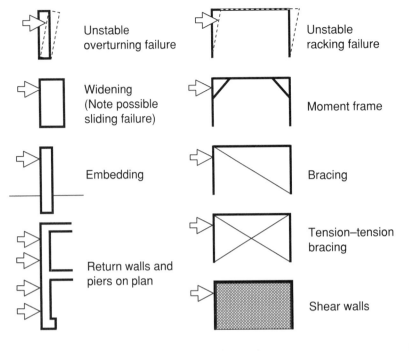

Figure 11.12 Stability of walls and frames.

(a) Wall stability options   (b) Frame stability options

in retaining walls resisting the horizontal forces from ground pressure rather than wind. They have limited application in the superstructure against wind. Walls above ground are almost certainly likely to be slender and will rely on return walls and piers (buttressing) plus the stiffening effect of floors. In frames the instability is racking. The frame can rely on moment (fixed) joints between the structural members, but this often is an expensive option. For frames with pin joints, bracing can be used, although it is possible to utilise other parts of the construction, such as internal walls or lift shafts as shear walls (core stability) or the external hull either braced or with moment resistance from the structure and/or façade. It is not necessary to stabilise every frame. As long as sufficient vertical section of frames is stabilised the whole building will be stable. The number and position of these stabilised vertical sections will vary, but structures must contain at least one section in order to stabilise each direction and at least one other section to resist a rotational

collapse of the whole building. An example of this will be given for a skeletal frame in Figure 11.15 later in the chapter.

Whatever additional support members are used these then become members with their own connections, distortions and internal forces. Consider the single diagonal bracing shown in Figure 11.12. Depending on wind direction it may be either a strut or a tie. It has to be designed as a strut as it is almost certainly slender and this makes it susceptible to buckling. The benefit of tension–tension bracing, as its name implies, is that the members only need be designed as ties, with only one tie acting at any one time depending on the wind direction.

Stress/strain and the choice of material

Visualising the sets of internal forces in structural members can be considered without any reference to the size, shape or materials of the structural members. However, it is an understanding of the type and magnitude of these distorted shapes and internal forces in the members and joints that gives clues to the design of shape and materials that might be possible within economic sizes and manufacturing process.

In order to proceed with the analysis it is necessary to propose the shape and size of the structural member. The internal forces can now be associated with the cross-sectional area of the member over which they will act. From this it is possible to determine the stress that will be generated in the material. The stress is determined by the internal force acting over the proposed area given by the size and shape of the cross section of the member. So far, the illustrations have shown solid cross sections. For beams this has been rectangular, but the distribution of stress across the section suggests an 'I' or box section might be used. This removes material from areas where the bending forces reduce towards the neutral axis, using less material and reducing weight, but it changes the stress distribution across the section. Whatever the cross-sectional shape the stress distribution is independent of the material chosen; it is only a function of size and shape.

The choice of size and shape, however, cannot be established without knowledge of the material's structural properties, mainly (although not exclusively) associated with the stress/strain relationship. It is now that it is possible to ask questions about the two key failures associated with structural performance:

- Will it break? This is known as the collapse criterion and is directly related to the strength characteristics of the material. For a safe solution the allowable stress in the material must be greater than the stress that will be generated by the internal forces over the area of material proposed.
- How much will it distort? This is known as the serviceability criterion and is related to the strain characteristics of the material. The distortion must be limited in order to maintain visual lines and ensure finishes do not crack or become detached. The structure must also not feel live (bouncy) when subject to the imposed loading.

In fact the simple question of whether it will break for the collapse criterion is not enough. In the interests of limiting the risk of collapse while designing an economic structure it would be more accurate to ask, 'How close is it to breaking under working conditions?' The design must incorporate a factor of safety. This will be greater the more sensitive the design is to variability of the loading and/or the materials properties. How this is allowed for depends on the approach taken to the design and is beyond the scope of this text.

For every material there is a relationship between the stress experienced and how much it strains, known as the e value. This relationship is shown in the stress/strain curves depicted in Figure 11.13 where the curves for concrete and

Figure 11.13 Stress/strain curves.

(a) Concrete   (b) Steel

steel are shown. The stress/strain relationship identifies the stiffness of the material, how much it strains with an increase in stress, which makes some materials more difficult to use in building to gain the full structural benefits. It may not be possible to take some materials to their full working stress without the strain induced by that stress leading to excessive distortions. Generally, building structures are made from relatively stiff materials such as masonry, timber, concrete and steel.

There is another important question in considering the collapse criterion. Not 'When will it break?' but 'How will it break?' This is another characteristic of the stress/strain relationship of the material that can be identified from the curves in Figure 11.13. It is the distinction between brittle and plastic failure. The curve for concrete shows that failure is sudden, it fails without any visual warning; it is a brittle material. The curve for steel shows that beyond a certain stress, known as the yield point, the rate of distortion is much greater as the stress increases so visual warning of failure is given; it is a plastic failure. A second characteristic is that releasing the load after the yield point has been passed will not return the material to its original shape, it is deforming plastically as opposed to elastically.

Once the stress/strain relationship of the material is known it is possible to evaluate the suggested shape, size and material specification for each member, ensuring that the stresses induced by the forces on the member are not excessive and the strains induced by the stress do not cause unacceptable distortions.

This process of checking suggested member size and shape against induced stresses and materials properties is the basis for structural analysis and design.

Structural analysis and design

Having established the loading pattern and devised a structural system of members and joints which is inherently stable, it is possible to identify the structural behaviour of each joint and member and determine the pattern of forces that will arise in each. Knowing the expected behaviour of the structure and materials under stress it is possible to suggest appropriate cross-sectional shapes, sizes, materials and details for the structure.

It is necessary to evaluate the performance of these suggestions to see whether they can offer sufficient resistance without breaking and within limits of distortion. Initial proposals can normally be made from observations of past construction combined with an understanding of structural theory and design methods. Theories of mechanics and strength of materials support engineering knowledge. These theories are used to devise mathematical relationships that predict forces, stresses and materials behaviour sufficiently accurately to be relied upon in design. They must, however, be based on an initial ability to visualise the behaviour of the structure and the materials involved. If the engineer does not imagine behaviour and then suspects that behaviour as being a possible failure mode, the calculations necessary to check the design may not be undertaken. Many structural failures can be attributed to this error of judgement.

Once behaviour of the stable arrangement of members is visualised and the most likely failure modes identified, calculating the magnitude and distribution of the internal forces induced in the members is known as the process of structural analysis. Choosing and evaluating the shape, dimensions and materials for the members that will safely resist these forces to limit distortion, local instabilities and ultimate loss of load carrying capacity is the process of structural design. Design also includes the detailing of the members and joints to ensure the behaviour assumed in the analysis.

Movements and structural behaviour

So far the analysis has assumed that the only 'movements' or change in shape or dimension from those originally formed are the distortion of the members under load. It has also been assumed that these will be able to take place freely, without any restraint from other parts of the building other than their connections one with another. In reality there will be other movements caused by changes in environmental conditions, which not only will change the dimensions of the structural members themselves but also may affect all the other parts of the building.

When these movements take place they may affect the structural system in one or more of the following ways:

- Restraining the natural distortions in the member itself will change the magnitude and distribution of the internal forces.
- If the member itself has to restrain movements from other parts of the building or the structural system, these will act as additional loading and hence change the magnitude and distribution of the internal forces.
- If movements elsewhere in the building result in a change in the position of some support in the system (most likely ground movements), major redistribution of the internal forces may occur (often determined by joint rather than member behaviour).

While every effort should be made to predict these movements and to detail the joints and connections to allow free movement, if this is not possible then the effects must be evaluated as additional loads and hence levels of stress.

An additional complication is that some materials, such as concrete, continue to distort with time even though the load remains constant. This phenomenon, known as creep, means that the effects of these movements may develop at some time in the future, leading to failure some time after the building has been occupied. A more comprehensive analysis of movements is covered in Chapter 12.

One aspect of movement always of importance in the design of a building structure is that of the relationship between any differential settlements in the foundations and the ability of a building structure to move with these settlements without distress. Each overall design of a superstructure (members, materials and joints) will have an associated flexibility that will determine how much differential settlement they can accommodate and this will be a key performance criterion for the foundation design.

Stresses in the ground

All the analysis so far has been to establish a stable arrangement of structural members, where the task has been to suggest the shape, dimensions and materials to be used and then check that the internal resistance is sufficient to carry the loads safely without distortion or breaking of any part of the structural system. The ground on which the building is founded is also part of the structural system but cannot be considered in this way. It too will have to be checked for stability (subsidence), distortion (settlement) and breaking or collapse failure. However, unlike any of the structural elements discussed so far, the ground cannot be specified, it has to be investigated.

The aim of the investigation is to determine the information necessary to check out the ground's ability to carry the loads without excessive settlement or collapse failure. However, it is not possible to determine the scope of the investigation or type of information required without some idea of the pattern of forces that will have to be resisted if the soil is not to fail. In this respect it is the same process as for all structural members, but the visualisation of the behaviour needs to identify the 'shape' of the stressed area. This is very different from the structural members considered so far since there is now no obvious edge or boundary to the soil as a structural member.

The two diagrams shown in Figure 11.14 help to show the way the soil will be stressed and explain both distortion (settlement) and collapse behaviour under a bearing foundation. To understand settlement, the diagram showing the bulbs of pressure (Figure 11.14a) gives an idea of how the stress will build up in the soil. This shows the volume of soil that will strain, leading to an ability to predict settlement.

There is a simple classification of soils into cohesive (broadly clays) and non-cohesive (broadly sands and gravels) which is useful when thinking about the behaviour of the ground.

(a) Compressive stress in soil leading to settlement   (b) Shear stress in soil leading to collapse

Figure 11.14 Stress in ground due to foundation loading.

One of these behavioural differences is how long it takes for the ultimate settlements to take place. In non-cohesive soils the settlements will take place as the loads are applied. Cohesive soils often take years to reach a final settlement and this time-dependent behaviour has to be investigated for a safe design to be achieved.

The observed collapse failure in soil shows how a shear plane develops in the soil which reaches the surface some distance from the foundation. Figure 11.14b shows two symmetrical shear planes. In reality only one needs to develop for failure, the foundation rotating with the one-sided slip. Comparing the two illustrations, pressure bulbs and shear planes, indicates that the volume of soil involved in collapse is not the same as the soil involved in settlement. Together the two diagrams show the extent of the investigation and the type of information required to predict the behaviour of the soil. The stress/strain (load/settlement) relationship is necessary to predict settlement and the shear strength needs to be known to check for collapse. These can be identified from laboratory tests on samples taken in the investigation or from in-situ tests taken in the undisturbed ground.

These direct limits of depth and spread of the area of investigation for settlement and collapse may miss the third failure mode, the disturbance or overall instability from subsidence. This would cause the movement of a mass of soil that removes the support for the volume of soil stressed by the loads from the foundations. Subsidence may be geological in origin with landslips or swallow holes or may be man-made with cut embankments or mining activity. This also has to be investigated, but the greater scope of area of investigation and the type of information required would probably need a desk study or topographical survey, although this may identify the need for a more widespread physical site investigation.

Another disturbance to the foundations not related to the direct stress in the soil is volume change due to changes in environmental conditions in the ground. Changes in moisture content in clay can significantly change the volume of the soil. In these 'shrinkable' clays, foundations must be taken below any risk of moisture change or the foundations may settle (or heave) independent of the stress in the soil from the building loads. Frost heave is another environmental change causing a volume change in the soil. Soils susceptible to movement under these conditions include silts, chalk, fine silty sands and some lean clay. Ground movement is discussed in greater detail in Chapter 12.

Other ground movements such as earthquakes generate stresses and changes in the soil properties, leading to new forces on the building, but these effects are beyond the scope of this text.

The major structural forms

It is now possible to return to the three major structural forms introduced in Chapter 3 and see the types of members and materials that might be used economically to achieve a variety of shapes and sizes of buildings for a variety of different uses.

Loadbearing wall structures

Structures based on loadbearing walls, by definition, use walls for support. In the past this has been the main structural form that has been developed for domestic-scale construction to larger buildings carrying high loads from relatively large spans, but this generated thick walls, increasing dead loads and limiting architectural form. The walls would have been based on masonry, probably stone or brick, or timber originally used to frame-up whole walls and more recently to frame panels that can be used to create walls of varying lengths. Developments in materials such as concrete and steel have provided other options in skeletal frames and long span roofs for these buildings. This has led to loadbearing walls now generally only being economic where floor spans are limited to say 4 or 5 metres. This limits loads, giving relatively thin walls. Where the use of the building requires relatively small internal spaces on a regular pattern that repeats on all floors so walls can be continued to the ground, loadbearing wall structures are an option. This is particularly true if the walls have sound or fire requirements, which tend to lead to construction that is good at carrying loads.

Modern materials have given stronger and more quality controlled products that have allowed buildings up to 18 to 20 storeys to be constructed but only where there are many internal as well as external walls available to carry loads. Typical of these buildings would be hotels, student residences and flats.

Given the limited internal spans, floors are likely to be concrete slabs, although in a single dwelling the timber joist (small beams at close centres) and boarding floor is probably the economic solution where sound and fire requirements are less. Roof structure is likely to be timber truss rafters (pin-jointed frame) on low-rise buildings.

If the cellular nature of the internal room layout provides walls in both directions the resulting structure is naturally stable against wind forces. If, however, the loadbearing walls are predominantly in only one direction on plan, the building will be unstable in the other direction and some other stability members will have to be introduced. These may be from internal shafts such as stairwells or might be short returns known as piers at the back and front of the building on the ends of the wall to stiffen the building in the vulnerable direction. Construction where major loadbearing walls are used only in one direction is known as cross-wall or box frame construction and allows a variety of front and back façades to be chosen for the building.

Skeletal frame structures

The development of concrete and steel as structural materials allowed the design of effective and economic bending members, particularly beams and slabs that form two of the major structural elements of the skeletal frame. Steel has high strength in both compression and tension. However, steel is expensive and to develop economic bending sections it has to be formed into the classic 'I' cross section. Concrete, while having no significant structural tensile strength, has sufficient compressive strength to make a composite with steel to resist the tensile forces, forming reinforced concrete members. The success of reinforced concrete as a composite is based on more than the relative strength characteristics of steel and concrete. The strain, thermal expansion properties and cost are also complementary. The shrinkage of concrete on curing, plus some ribbing on high yield reinforcement, provides the bond necessary for the two materials to act as a composite.

With the newly developed high-strength materials comes the ability to take vertical loads on small cross sections so walls could become columns. The basic form of the skeletal frame

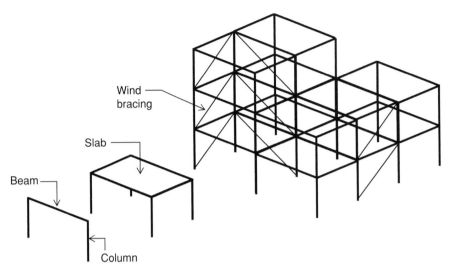

Figure 11.15 Basic members and arrangements of skeletal frames.

shown in Figure 11.15 could now be formed effectively to offer not only more clear internal space in buildings but also greater height without the greater thickness of materials that went with the limited strength of masonry. Floor slabs are the third major component of the skeletal frame. Almost universally constructed in reinforced concrete, even for structural steel frames, these floor slabs have been developed in a number of forms both in situ and precast. This gives a range of options that can accommodate spans between 3 and 12 metres, but their economy is determined by their choice as part of the frame layout as a whole.

It is not normally economic to have columns at much less than 6 metre or more than 12 metre centres. This is linked to the choice of slab. This would mean, for instance, the use of secondary beams for a 3 metre one-way spanning slab or up to a 12 × 12 metre column grid if a two-way spanning waffle slab were used. Slab developments have also led to reinforced concrete frames without beams where plate or flat slab floors have slabs transferring their loads directly to the columns. This generates high punching shear stresses around the columns. These floors with office loading can span 9 metres at around 350 mm deep in reinforced concrete and even greater spans (or shallower depths) if pre-

stressing is employed. The economy in the flat slab is achieved in the simplicity of the formwork and falsework systems which are a major proportion of the cost of reinforced concrete structures.

Wind stability members make up the other major structural member of the skeletal frame. Shear walls are probably the natural choice for in-situ reinforced concrete frames. Compression or tension—tension bracing may be used with steel or precast concrete frames, although concrete or masonry cores are often an alternative, depending on the layout of the buildings.

The skeletal frame no longer requires external walls to carry dead and imposed loads from the building, only to transfer wind loads to the structure. This has allowed a wide variety of façade materials to be employed to give expression to a range of architectural forms.

Long span roof structures

The need to roof over large areas with little interruption even from columns characterises a number of buildings that are commissioned. Initially developed for warehouses and factories, this form is now employed for retail and sports and leisure use. Predominantly used to cover ground floor areas, they can also be used to form a roof over upper floors where the activities on lower floors can be accommodated in a skeletal frame structure in buildings such as airport terminals.

Although 'long' span roofs are often thought of as over 35 metres span, the simpler plane frame, and even some of the three-dimensional forms, can be economically used over much shorter spans. Although not universally accepted as a definition, it is useful to think of these forms as being used on short spans of up to around 15 metres and as medium span from 15 to 35 metres. This gives some clues as to which forms might be economic.

While, like skeletal structures, they are based on the frame (a bending structure supported on columns), they have been developed into a much wider variety of forms. Many are expensive and gain their value in their dramatic forms over medium to long spans and are often left exposed as part of the architectural design.

The simplest and generally least expensive form is the plane frame. There are now two common forms: the pin-jointed frame, such as trusses and girders supported on columns, and the portal or moment frame, where fixed joints distribute the bending moments into the columns, giving economic 'beam' depths even over large spans. The triangular-shaped truss used to form a pitched roof is only normally used for short spans, but both girders and portals can be used for a long span of up to 100 metres if required, although smaller spans are more economic.

The spacing between the frames will be dependent on the ability to provide purlins and cladding rails to support the cladding (see Figure 11.7). On longer spans the individual frames become more expensive so setting them further apart can reduce the cost of the frames, but the purlins and cladding rails become more expensive. On short to medium spans, frame spacing of 4 to 8 metres with cold rolled steel Z purlin and cladding rails might be used. On long spans, 8 to 12 metre spacing of frames with light truss purlin and cladding rails may become more economic.

Wind stability provision is less straightforward than for skeletal frames. Portal or moment frames are inherently stable against side wind which will put additional bending forces in the frame (sway) that have to be analysed. Being plane frames and not connected together other than by purlins and cladding rails they are unstable against end wind loading and require bracing in this direction. The long span creates large roof areas and the light roof cladding makes these frames susceptible to wind uplift. Given that the roof sheeting remains connected to the structure, the frame may be subject to bending reversal. The top of the beam will go into tension and the bottom is subject to compressive forces. In these frames one failure may become lateral torsion buckling, as the bottom of the beam is unrestrained. In steel portals, for example, struts have to be specified to the bottom flange of the 'I' beam against this condition.

Pin-jointed trusses and girders, with pin joints to the top of the column, are vulnerable to wind in both side and end loading. The end-loading wind may cause the frames to overturn not only at the connection at the bottom of the column to the foundation but also at the connection of the truss to the top of the column. Bracing for the walls and at the eaves will be necessary.

As the height to eaves of these frames and/or the spacing apart becomes greater the length of the bracing member becomes longer. If single bracing is used, the increasing slenderness makes the buckling failure more likely and large-section members have to be used. If tension—tension bracing is used the increased length does not increase the chance of failure, making the two-member form of bracing economic.

The second broad type of long span roof is the three-dimensional form. Either it takes advantage of the two-way spanning pin-jointed frame forming flat, two-layer grids such as space frames, or it gains the spanning advantage by curving (or folding). Curving can be achieved either in single-layer grids such as geodesic domes or from solid slab-like structures such as shells or folded plates. These forms are relatively stiff, but curving can be used in the more flexible tension structures based on suspended nets (cables) or fabric structures.

A full analysis of the spanning capabilities and characteristics of each of these three dimensional forms is beyond this text. They can all be used over medium and long spans in a variety of construction options and materials, achieving a wide range of architectural forms, making the analysis of their choice complex, with a need to be well informed. It is not sensible to offer broad advice on form and spacing of these three-dimensional forms as it is with the skeletal frames and the plane frame solutions for these long span roof structures.

Summary

(1) To consider structural behaviour it is first necessary to understand the external loading conditions, primarily dead, imposed and wind loading, although they may include impact and moving loads.

(2) While all parts of construction are subject to at least some of these loads and therefore have to be considered for strength, damage resistance and adequate methods of fixing, there is for all buildings a recognisable structural system transferring these loads to the foundations. This system will comprise a series of structural members and supports jointed together to provide a load path through the building.

(3) Five basic structural members can be identified: the strut, tie, beam, wall and slab. Each can be used for a range of loading conditions and support arrangements which will determine the distorted shape and will give clues to the distribution of internal forces that will be generated. These forces are compression, tension, bending shear and torsion. Closer analysis of these forces shows them all to resolve into either compression or tension.

(4) In addition to visualising the distorted shape that arises in the direction of loads, lateral distortions will appear in slender members in compression at right angles to the direction of loads. Eccentric loads will also introduce new distortions, leading to new internal forces.

(5) Curving spanning members can be used to advantage. Curving in the direction of the span can eliminate bending forces, leaving the member in compression (arching) or tension (sagging), although this leaves horizontal forces at the supports. Curving at right angles to the span increases the depth, giving a more effective bending section with less material.

(6) The way members are connected also affects the distribution of internal forces. Pin joints, which allow rotation, do not transfer bending, while fixed or rigid joints do redistribute bending stresses into the support.

(7) The pin joint can be used to build up spanning members into grid structures such as trusses and girders which become a framework of struts and ties, making large, relatively light, spanning frames for longer spans.

(8) When members are arranged into building structures it is possible to see the way in which wind loads are applied. It is likely that the structure visualised to carry the dead and imposed vertical loads will be unstable against wind loads, and other structural members may have to be added. The forces in these new members will have to be analysed and the forces in the original members reassessed.

(9) With the whole structure stabilised and the pattern of external and internal forces established, the shape, size and materials for each member can be specified. The shape and size of the cross section of the member will determine the area over which the internal forces will act. This will give the stress that will be experienced by the material. Knowledge of the stress and strain characteristics of the material can be used to check

the suggested shape and size for sufficient resistance to keep suspected failure within factors of safety.

(10) All this analysis assumes that movements in the building do not cause a transfer of load from or to either structural or non-structural parts of the building.

(11) The ground is part of the structural system of the building but cannot be defined and specified in the way other structural members have been considered. It has to be investigated and the scope of the investigation is determined by the way stresses appear in soil causing settlement and collapse.

(12) It is now possible to return to the three major structural forms – loadbearing walls, skeletal frames and long span roof structure – to identify the economic arrangements, sizes and materials that might be chosen for a building structure.

12 Physical Behaviour Over Time

This chapter deals with the third major aspect of the physical behaviour of the construction: consideration of the analysis of deterioration, movement and wear. All these processes will happen during the life of the building and may reduce the reliability of the construction to continue its contribution to the performance of the building. Concerned with the changes that might take place with time, it considers the potential failures and the implications for maintenance and repair during the life of the building.

Reliability: renewal, maintenance, repair and disposal

The analysis of behaviour over time is about understanding possible changes that take place in the materials, components, joints and fixings that will affect the ability of the construction to perform. Changes may reduce the chance of performance the next time design conditions are experienced. The reliability of the construction is reducing. Its chance of failure under the next set of testing conditions becomes greater.

However, in most cases, construction is not just built and then demolished at the end of the life of the building. Its reliability may be dependent on interventions during its life to keep satisfactory performance. To evaluate a proposal against its reliability over time it is necessary to specify:

- How long it is intended to last before it is renewed
- What maintenance it can expect during its life
- What is the impact of failure and the ease of repair

To complete the specification it is then necessary to identify:

- How it can be dismantled
- What options there are for safe disposal

Without both the initial detailing and specification and a clear statement on these interventions in the life cycle of the construction, the proposal has not been fully evaluated.

Not all parts of the building have to last for the same length of time before renewal. Clients' and users' requirements change and, while not foreseeable in detail, they are likely to involve finishes and the division of space more than structure. It may be more economic to renew than to maintain, although this may not be desirable against sustainability criteria. It is necessary to establish as part of the analysis the client's commitment to renewal cycles and maintenance plans.

Another strategy for some parts of the building is to expect failure and renew or repair if and when this happens. For the client this has a cost burden for loss of use and in establishing an organisation of repair in the case of failure during the operation of the building.

These decisions on repair and maintenance concern all parts of the building from tap washers through decorations and roof finishes to the life cycle of the building itself, including expansion and change of use. Clients will have a view on the management of the facilities they use for their everyday operations. Decisions on value for money and sustainability are not just concerned with initial costs but the whole life cycle of the construction.

The basis of choice is therefore to understand the changes that take place over time and then consider the options for renewal, maintenance and repair to keep operational performance. Proposed construction will have to have an initial specification that should have foreseen the anticipated maintenance, with an analysis of the consequences of failure (safety, cost and disruption) and the ease of repair, with requirements for access that have to be incorporated into the design. If the client does not favour maintenance, and repairs cannot be made easily, then initial specification and detailing may have to be changed before the final choice is made.

Finally there is the consideration of dismantling and disposal which may also include decommissioning. Whether considering renewing a part or the demolition of the whole building, the ease with which it can be taken apart and the impact of disposal should be anticipated in the original specification and details. When this is seen as an economic issue it may be considered to be so far into the future that it can be ignored at the stage of initial design. However, with the increasing emphasis on sustainability with its focus on the quality of environment for future generations, dismantling and disposal can no longer be so easily dismissed at the design stage.

Basis of analysis

In the past the limited number of construction materials coupled with relatively low performance expectations allowed knowledge to build up over the years. Details were developed to protect parts from decay. Where protection was not possible, there was a greater acceptance of maintenance activities to keep the construction in good condition. The details that developed were often regional, if not local, to fully take into account the particular conditions for the locality. More than the detailing, the orientation and grouping of buildings took into account local exposure conditions, further protecting the vulnerable parts of the building from the worst ravages of the weather.

This description of the close links with experience, locality and building traditions no longer holds true for the majority of buildings erected today. This places greater emphasis on the designer's ability to undertake an analysis of behaviour of the construction with the passage of time.

There are three broad areas of analysis that have to be considered as significant to the reliability of building, each with a potential to lead to failure, invoking the need for maintenance, replacement or repair:

(1) Environmental deterioration (durability)
(2) Movement
(3) Wear

There is a need to understand their direct effects in order to undertake an analysis of the proposed solution to identify ways in which the building might become less reliable, require effort to maintain, and ultimately fail.

Durability of materials

Over time, changes take place in the properties and integrity of the materials of which components are made. This will alter their performance in the future. It should be acknowledged that some of these changes are beneficial in that they improve the durability of the materials. The weathered surface of stonework is a good example of this. This makes the material more durable. It is important to recognise this not only when specifying new building work but also when considering cleaning, maintaining or altering the building. However, many changes that take place reduce the durability of the material.

Durability of a material is dependent on the conditions to which it is subjected. Each material has specific agents of degradation or preservation. It is not possible to determine the durability of a material in a building without a clear definition of the conditions under which it will have to operate. Each material has associated agents and mechanisms that are known to degrade or preserve the materials. Either material must be chosen to survive in the expected conditions or, unless a new material is specified, some protection or preservation must be specified.

Agents and mechanisms

The mechanisms of change can be identified as taking place either within the material or at the surface. In both cases there will be a change to the internal structure of the material that is affected. This internal change can either affect the properties of the material or disrupt the internal integrity of the material, both of which result in a change in performance, most often for the worst. Changes in properties include embrittlement in plastics and fatigue in metals, while examples of internal disruption are wet rot in timber and sulphate attack in ordinary Portland cement-based materials. In each of these examples changes only occur in the presence of specific agents. Embrittlement is associated with ultra-violet light, fatigue with reversing stresses, wet rot with dampness and sulphate attack with sulphates and moisture.

These examples also serve to bring out another useful way of thinking about these agents of change. They are chemical (sulphate attack), biological (wet rot) or physical (fatigue). There is often a coincidence of conditions as in frost disruption where saturation and cold temperatures cause a change of state, not in the materials, but in the saturation water, causing a physical internal disruption of the material normally just below the surface.

Study of the major materials used in building shows that water plays a part in many of the deterioration mechanisms, as does air with its ready supply of oxygen. These components are not only significant in many biological processes but also in chemical processes, for example oxidation.

This is not perhaps surprising, for the processes of nature are dependent on decay and regeneration. Many of the materials from which we make our buildings are taken from nature and they are, therefore, part of the decay–regeneration cycle. Understanding the mechanisms and agents in this cycle becomes the key to predicting the durability of many materials and diagnosing failure.

Many of the more recently adopted materials we use are not naturally occurring but have been refined or processed. The pathways and agents of change in these cases are often to revert to their original or lower-energy states. Many processed materials have proved to be stable with time, for example glass. However, there is a tendency in some processed materials to change to lower-energy states which may not have the properties that we carefully designed into the processed materials. Many of these mechanisms also need agents, again often water and oxygen, with other chemical agents accelerating the process. The mechanism of oxidation in metals is a good example of this.

Oxidation is a surface process where the oxidised layer does not have the properties of the parent material. Oxidation of iron (rust) is recognised as reducing the life of a component. Not only does the oxidised material have a greater volume than the parent material, which increases the dimensions of the original component, but also the rust layer itself ruptures and allows the agents of oxidation to reach down to new parent material, continuing the cycle of degradation until all the material has rusted away. However, this is another example where not all changes reduce durability. Non-ferrous metals, such as aluminium and copper, oxidise, but this new surface, while of greater volume, changing the dimensions of the component, is not great enough to rupture the surface layer itself. This new oxidised layer or 'patina' is stable at normal temperatures and therefore protects the underlying parent material by significantly slowing down the rate of degradation. The component's life is potentially increased.

The identification of agents associated with the mechanism of deterioration or preservation for each material is a key aspect of the analysis of the durability of a material. It involves the identification of the environment in which the material is to function.

Cycles and concentrations

Because the rate at which changes occur in materials varies with exposure to the agents of change it is necessary to project an expected pattern of exposure. The pattern must identify the frequency with which changes in concentrations of agents, which are in contact with the materials, are expected to occur. Because many deterioration mechanisms have a number of agents involved, a pattern for each has to be established so the frequency of coincidence can be identified. In frost damage, not only is the coincidence of saturation and low temperatures important, but also the frequency of the freeze–thaw cycle is crucial to the rate of deterioration. If the material remains dry at low temperatures there is no deterioration. If the material is saturated but remains frozen for long periods there is little chance of deterioration. If the material is saturated and freezes at night and thaws during the day, there is a greater chance of deterioration.

It is very important to carry out this analysis of conditions very locally, in immediate contact with the material. It is very possible that even different parts of the same component are subject to very different patterns of exposure. Failure is often very localised, as with the wooden fence post rotting around ground level shown as a high-risk area in Figure 12.1. Here the patterns of concentrations of moisture, air and decay organisms are at their greatest and the cycles of change at their most frequent. This is also a good example to identify the preserving nature

of some agents if their concentrations remain high but steady. The example here is water and timber. If water is constantly present, maintaining the timber in a saturated condition, it acts to preserve. The fence post buried deep in permanently wet ground does not decay. To continue the fence post example, the top of the post is also vulnerable as it is end grain that allows the water deeper into the timber, providing a pathway to high local concentrations of the decay agents.

This pattern of conditions allows an analysis of the rate of deterioration that takes place at the critical point in a component. Most materials in the building will have periods of decay and periods when little change is taking place. This may, however, in most cases be considered in the long run to be a gradual change. The changes taking place are making the component less reliable the next time the component has to operate at or near its design limits. The fence post breaks in a high wind but no worse than winds it has survived in the past.

Figure 12.1   Timber fence post – deterioration and improved durability specification.

Where deterioration patterns are identified as creating a high probability of failure (or the consequences of failure are catastrophic) then consideration has to be given to changes in the suggested specification. It may be possible to protect the vulnerable parts from the agents of decay or provide some preserving process that limits the effects of the decay agents. It may be economic to change the specification of the material. In the case of the fence post the end grain at the top can be protected with a cap, the timber preserved particularly where it enters the ground or a metal shoe placed in the ground to receive the timber above the aggressive line of the ground. In this case the deterioration of the metal shoe would have to be analysed.

The maintenance expectations for this specification of the fence post might include the gradual reduction of the effect of the preservative, which would have to be replenished, applied by brush some time in the future. Maintaining and/or repairing of the cap at the top of the post would be cost effective as it is easy to do. The ultimate failure may well now still be at the base of the post even with the metal shoe. If maintenance of the metal shoe is not contemplated then it may be that repair of the first few posts to fail could prolong the life of the fence, but ultimately the renewal of the whole fence would be the preferred option. Dismantling would not be problematic and neither would be the disposal of the timber as long as the preservative was not still active. It may be that at this stage the metal shoes may be cleaned, renovated and reused.

Movements in components

The term movement here refers to changes not only in position but also more generally in size or shape of the components. All buildings and their components will be subject to these movements, but, if correctly detailed, this need not lead to failure. Movements of this kind can be limited, accommodated and allowed for but cannot be eliminated. The dimensional changes are often termed inherent deviations as they occur due to the inherent properties of the materials. This is in contrast to the induced deviations that are the result of the production process and need to be considered to ensure that the initial construction will fit together, as discussed in Chapter 4.

Many movements pose little threat to the building. If they can take place freely, without impacting on adjacent construction, they will have little effect. Movements that are of concern are generally differential, where one part is moving more than another to which it is connected, or restrained, where one part is moving in one direction and meets another that cannot move (or is indeed moving to meet the first part). However, some movements need be very small to have a large effect. The shrinkage of cement mortars in brickwork breaks the bond with the brick, causing fine capillary paths leading to the potential for damp in the building, as discussed in Chapter 10, where the most common solution is to introduce a cavity to break the damp path to the inside of the building.

Much of this movement can be accommodated in the normal detailing as long as care is taken in thinking through the probability that induced deviations will not eliminate the gaps and/or that the limit of flexibility of jointing materials that allow the movements to take place freely is not exceeded. These details then have to be built with care to ensure that the expected allowance is achieved in reality. Some movements are so significant to the integrity of the construction that specific movement joints have to be designed. These may be required where structures of different heights come together with different foundation settlements. Another example is the design of large panels of brickwork where environmental changes will cause internal stress in continuous construction.

The origins of these movements can be traced to one or more of the following:

- Initial equalisation immediately after installation
- The application of loads
- Changes in environmental conditions
- Volume change associated with deterioration
- Ground subsidence

Initial equalisation movements

Many materials undergo dimensional changes early in their life while the material is developing its final properties, often through changes in temperature or exchanging moisture. These movements are normally irreversible. In processing and manufacture, many materials have to be prepared to make them workable to produce the shapes and forms required in the component. For some components made off site, at least some of these equalisation changes in dimensions take place before they are delivered. Many components, however, have not completed this change when built into the building, and for work carried out in situ all the dimensional change takes place in the final position.

Materials most often built into the building while this change in dimension is still taking place include timber, concrete and clay products such as bricks. Timber components are manufactured at moisture contents above those that will eventually be achieved in the occupied building and therefore will continue to shrink after installation. The equalisation movements in concretes, plasters and brick also include changes in moisture but in different directions. Concretes and plasters need water for setting and workability and then dry out and, therefore, like timber, shrink. Bricks that are fired and therefore very dry at the end of the manufacturing process take in moisture and therefore expand after the initial cooling shrinkage has taken place.

The important factor is often the timing of the installation. Concrete blocks laid too soon after casting will continue to shrink, causing tensile forces leading to cracking in the wall. While the majority of expansion in bricks takes place in the early weeks after firing, they continue to expand well into the initial life of the building. Some continuing expansion of brickwork should be anticipated. As brickwork normally has sufficient internal compressive strength this expansion will lead to the whole panel sliding on the damp-proof course (DPC) or cracking at short returns.

In in-situ work, all the initial equalisation takes place in position. Drying shrinkage in concrete laid in large areas such as ground floors can lead to cracking due to internal tensile forces that build up during the equalisation phase.

Loading and environmental movements

These movements are dependent on the properties of the materials and conditions experienced by those materials during the life of the building. Movements caused by loads will depend on the stress/strain relationship of the materials and on the distribution and magnitude of the loads themselves. The two major environmental agents are temperature and moisture. Nearly all materials change dimension with change in temperature. The amount of movement that takes place in a material for 1°C change in its temperature is its coefficient of linear expansion. Only permeable materials will be subject to moisture movement. Like temperature changes, the change has to take place within the material, not just in the external conditions. The amount of moisture movement that can be expected is normally quoted as a percentage of the original size.

The timing of movements caused by environmental and loading conditions can be seen to be associated with two phases that may overlap in time. The first phase takes place predominantly during construction and in the early stages of occupation, where the movements are usually not reversible. These are taking place at the same time as many of the equalisation processes and are initially associated with loading. Structural members are gradually loaded (dead and imposed) as the building is constructed and during the initial phase of occupation. This loading increases stresses that increase strains, causing the distortions discussed in Chapter 11.

In addition, the temperature and moisture conditions can be very different during construction depending predominantly on the time of year. This determines not only how quickly equalisation processes take place but also the size of components at the time when adjacent construction is installed.

Once the building is occupied most of the loading is completed, but the internal conditions are now being established for the first time. This will move the temperature and moisture to working conditions and during this period more dimensional changes will be taking place. Not all strains from the applied loads take place immediately. While settlements in non-cohesive soils will take place as the loads are applied, consolidation in clay soils may take years to complete, meaning that not all foundation settlement is experienced in the initial loading period. Some materials, including concrete and timber, exhibit creep. This is a continuing strain under constant sustained load. The rate of strain decreases with time but can manifest problems some years after the construction of the building. This period could cover the first few years of occupation.

The second phase in which movements may take place covers the rest of the life of the building, after the first few years of occupation. These movements are usually reversible. Most of the equalisation process will have taken place.

The effects of environmental changes in moisture and temperature on external materials become significant in this second phase. Exposed areas can experience wide variations in temperature. In temperate climates these are not so acute on a daily basis but are experienced in the annual cycle of the seasons. While air temperature varies from summer to winter, materials exposed to the radiant effects of the clear sky experience the greatest surface variation. Midday summer sun can raise the temperature of surfaces, particular if dark in colour, to temperatures of 60°C. Conversely, clear night skies in winter will reduce temperatures of surfaces directly exposed to the sky to −20°C, again particularly if dark in colour. This potential 80°C difference will significantly alter the dimensions of many of the common materials used externally on buildings. Particularly vulnerable are continuous roof coverings, leading to the specification of white-coloured protection to limit the temperatures experienced by the roof covering itself.

External materials also experience wide changes in moisture, from prolonged periods of sunshine to extended periods of rain. Perhaps this is most often noticed in timber doors that 'stick' in wet weather. The materials affected, which have to be permeable to absorb the moisture,

fall into two categories: rigid internal structures and elastic internal structures. These are broadly ceramics such as brick and concrete and organic materials, predominantly timber, respectively. Ceramics are subject to relatively small changes, while timber is subject to significantly larger changes. Timber shrinks and expands differently in different direction to the grain, changes being greatest across the grain.

Interior components are subject to less variation in temperature and moisture unless the activity undertaken in the building creates variations in these environmental conditions and no specific measures such as ventilation are taken to limit the areas affected. It should not, however, be assumed that all internal components will remain stable. Windowsills are particularly vulnerable to sunshine through the glass, raising their temperature, and fittings associated with heating systems will be subject to wide changes in temperature. Pipes should be able to move freely where they pass through the construction or they may 'creak' as the movement takes place.

Movements due to temperature and moisture changes only take place if the changes occur within the material, not just at the surface. However, surface changes can create internal stress as the surface tries to change but is restrained by the mass of the material. This image of layers of construction expanding and contracting at different rates is particularly important if layers of different materials are to be bonded together to act as a composite. If thermal expansion properties are significantly different then changes in temperature from those at the time of bonding will create bending stresses in the materials and shear stresses along the bonded surfaces. This is also true for moisture changes. Materials bonded together that get wet can suffer similar stresses to those suggested by temperature changes if their moisture movement characteristics are significantly different.

Volume change associated with deterioration

In considering durability earlier in this chapter, some of the changes were in the properties of the materials, but some involved changes in volume. They were mainly associated with the chemical changes, such as sulphate attack or corrosion, but would also include frost disruption. These changes in volume not only create the potential to disrupt the component itself but also have a potential to create additional stress in adjacent components with which they are in contact. An example of this is the attaching of stonework using iron fixings that rust.

Ground movement, settlement and subsidence

The ground is subject to two types of movement: settlement and subsidence. Settlement is the result of loading and moisture changes in the soil and therefore its analysis is the same as that for loading and environmental changes of any other part of the building.

In assessing the consequences of settlement it is important to remember that it is the overall flexibility of a building that determines the amount of settlement it can sustain without distress. Brick structures in cement mortar can take less differential settlement without cracking than brick in lime mortar. Frames with pin joints can take more differential settlement than continuous structures without a redistribution of stress that may cause failure.

To assess settlement, soils are often characterised as cohesive and non-cohesive, representing the clays and the granular soils respectively. This is a useful distinction when considering movements in the soil carrying the building. Under loading conditions, clay takes considerable time (sometimes years) to compress (the action is consolidation) and even then may be subject to creep. Granular soils compress (predominant action compaction) almost immediately the load is applied and do not change unless the loading is changed.

Clay soils undergo high volume changes with changes in moisture, not a characteristic of granular soils. Processes of desiccation in clays include seasonal drying of soil in the summer and removal of moisture by vegetation, particularly trees. These movements can be very large and very powerful. Not only will a drying and therefore shrinking clay cause additional settlement but also clay taking up moisture will heave with sufficient power to lift the building. Generally, foundations should be established sufficiently deep to be below any chance of these movements affecting the support of the building. Piles may need to be sleeved as heaving clay will grip the pile and lift the whole pile with the building on top of it.

The freezing of water within the structure of the soil can also cause heave in some soils, typically silts, chalk, fine silty sands and some lean clay. Foundations should, therefore, be placed below the frost line in these soils (unless building on permafrost, a condition found in some cold regions of the world).

Increasingly buildings are being built on ground that is neither cohesive nor granular as defined by geologically laid down material. Landfill, be it from excavation, demolition or domestic waste, is creating new development sites. This material is highly variable and cannot be easily generalised. It may have been well laid with some measure of compaction or it may have been just tipped. The engineering properties of such materials have to be carefully investigated before foundation decisions are taken. Taking deep foundations through the fill, 'floating' the building on a raft or improving the ground to allow shallow foundation in the fill may represent the economic answers. This will depend not only on ground conditions but also on the building structure, taking into account the scale of development and the flexibility of the construction.

Subsidence causing a loss of support to the soil acting as the foundations can occur with geological changes or from human activity. Geological changes can be deep, such as earthquakes, which, while a real threat in some parts of the world, are not a design consideration in many others. Geological processes can also be surface actions, with slope stability being a particular threat to buildings. Natural slopes can become unstable in extreme weather conditions, but man-made slopes created as part of the development process can also become unstable, threatening the support to foundations. If slopes have to be steep, then retaining walls may have to be built to protect the buildings at both the top and bottom of the slope.

Mining activity gives a particular subsidence problem. When some mine workings are abandoned they may collapse, causing a lowering of the ground surface in a wave behind the collapsing galleries of the mine. Buildings that exist as the wave passes are vulnerable, but after it has passed, new development is less threatened. Shafts and swallow holes (natural shafts which open up in limestone areas) can also pose a threat and site investigation and a search of the geological and mining records are essential in areas known to be active.

Movement and detailing

Movements occur because of changes in shape, size or position of the materials and components that may or may not in themselves create any damage. The possibility for damage often lies in the detailing; the way the parts are connected and fixed. As soon as parts are connected or touching then there is the potential for the movement or its effects to be transferred. If the detailing allows the movement to take place freely then only the component itself need be considered. If no such freedom can be identified in the detailing then the movement will be restrained and force will develop between the moving and restraining components. This force may move adjacent components or may induce stresses that distort parts that are still held in position; finally it may induce stresses sufficient to crack or break components or the fixing between materials.

Estimating possible movements in a particular material or component is, therefore, only the start of any analysis. It becomes necessary to be clear on the details to identify free and restrained movements.

The first part of the analysis of details involves estimating the possible position of the parts when incorporated into the building. This cannot be assumed to be the position shown on the drawing. Each operation on site can only be achieved within certain dimensional limits associated with induced deviation. The exact position within these limits cannot be known at the time of specification. It can only be assumed that construction will be achieved somewhere within the extremes of these allowable tolerances.

Movements take place from this initial installed position. Equalisation changes that have not taken place at the time of installation will cause movement, loading as it is applied will cause movement, environmental changes may induce movements and later deterioration may create a change in dimensions, again causing movement. Not all of these will always be cumulative. Expansions may tend to cancel out shrinkage, as with concrete where operational thermal expansions are really greater than equalisation drying shrinkages. The shrinkage in concrete may, however, aggravate other details such as in the connection of a clay brick façade, which will tend to expand while the concrete frame will almost certainly shorten by both concrete curing and subsequent loading (including creep). The form of a movement joint to accommodate these movements while remaining weatherproof is shown in Figure 12.2.

If this analysis predicts failure then modification needs to be made to the specification or the detailing. Materials with different properties may have to be considered. Protection for the conditions creating the movements could be considered. However, the most appropriate option may be the redesign of joints or connections to accommodate the movement, allowing it to take place in a controlled way where the detailing can include gaps or materials capable of taking the movements without loss of performance.

Wear of components

The section on movement has concentrated on the changes in dimension, shape and position that occur in components when loading and environmental conditions change. Wear is also

Figure 12.2   Brick façade on reinforced concrete frame – detail of movement joints.

the result of movement, but in this case in parts that are designed to move, such as hinges or moveable partition tracks. Wear is not determined by the conditions to which the components are exposed but by the frequency of operation. It is only a small part of the analysis of most construction because the majority of building construction is concerned with passive performance.

Wear is more of a consideration in services where moving parts are often associated with motors and machines that are part of services systems. Here the possibility of loss of performance at some future date requires access to remove and replace parts or even replace the whole unit. This has to be considered in the design of the building as well as the design of the parts themselves.

Summary

(1) During the life of the building, changes occur in the materials and components that affect the reliability of the construction to perform the next time it is tested under its design conditions.

(2) The three broad changes that might bring about a loss of reliability and failure are environmental deterioration (durability), movements and wear.

(3) Environmental agents causing deterioration are specific to each material, and therefore it is necessary to determine the concentration and frequency of these agents in the environments associated with the specific components of the building, often very localised to one vulnerable part of the detail.

(4) Movements, or changes in shape, size or position of components, can be caused by initial equalisation processes following installation as well as the loading and environmental conditions of temperature and moisture changes that may take place during the life of the building. Movements in the ground then require a separate understanding associated with subsidence, which is in addition to the loading and moisture changes that cause settlement.

(5) Wear takes place in parts designed to move and is, therefore, more significant in services design where provision has to be made for replacement of parts or whole units.

(6) Not all these changes lead to failure. Distress in the building leading to loss of performance will depend on the detailing limiting deterioration and accommodating movements.

13 Manufacture and Assembly

This chapter considers an approach to the analysis of the production process involving both manufacture and assembly. It suggests that it is not just the concern of the production manager but needs to be understood as an integral part of choosing an appropriate construction solution. Starting from the need to ensure performance it is necessary to visualise the construction in pieces and then at its various stages of partial assembly. The work required to complete each stage has to be seen as a series of operations achieved through chosen methods that determine the mix of resources necessary. The resources of materials and labour are supported by two categories of production equipment, temporary works and plant and machinery. The methods chosen must ensure a safe, economic and timely completion to an appropriate quality to ensure performance. When the analysis shows that these can be achieved the suggested solution can be said to be buildable.

Realisation of performance

In considering the physical behaviour of construction in Chapters 9 to 12, the analysis has taken the completed building to be under operational conditions. The proposal represents a potential performance: only the final building can realise an actual performance. It must, therefore, be part of the evaluation to question whether it can be built at reasonable cost (economic and environmental), in a reasonable time and in a safe manner and achieve the performance potential.

The purpose of the analysis in previous chapters has been to specify and detail the materials, shape, dimensions and relative positions of the parts of the completed construction. Any analysis of production needs this information in order to think about methods and resources that might bring these requirements to reality. This does not mean that it is impossible to think about production until the solution is finalised. Even if the solution is at an early stage, with only broad ideas of possible materials and dimensions, the early indications of production options are established. It is possible to recognise production implications as soon as a physical form is emerging. This may then influence the choice of construction as the solutions become refined.

Visualising stages and sequence

The analysis of the behaviour of the final construction, which has been considered in previous chapters, has required the ability to visualise the completed works. It has been necessary to be able to see the construction fully assembled and under working conditions. To consider either manufacturing or assembly processes it is necessary to be able to visualise the building in pieces, to be able to see the materials and components at various stages of manufacture, to see how each part might look when it arrives on site and how it has to be prepared for incorporation into the final assembly. This then leads to the ability to visualise how the building will look at each stage of construction, partially erected. This is necessary to analyse not only the methods and resources of production but also the behaviour of the partly assembled building to ensure it remains in a safe condition at each stage of production.

The production sequence has to be visualised as a series of stages, each being a recognisable safe condition. It is then necessary to identify the operations that have to be performed to move from one condition to the next. These operations can often be undertaken in a variety of ways, each with their own mix of resources that have to be of sufficient quality and be available in sufficient quantity. Being able to visualise an appropriate method to produce the building safely, economically and to achieve performance is the criterion in declaring the construction to be buildable.

The basis of the analysis of buildability is to be able to identify all the operations that are necessary to complete each stage of the construction. A very large range of operations is involved, from actual handling, preparing and assembling of the parts to technical and managerial control. Each of the operations has to be identified and a method chosen that will determine the resources necessary to achieve the next stage of production.

It is necessary to be able to see these stages, identify the operations and decide on methods before it is possible to check that the resources can be obtained to complete the works.

Manufacture and assembly

The overall process of production can be characterised as taking processed materials, forming them into components and then assembling them to create the building. Among the primary decisions to be taken in this process is where these activities should take place. In an industrialised society there are more options available as to where the components can be made and assembly processes undertaken.

In 'traditional' construction the components are created and assembled by craft skills. This involves cutting and fitting on site, leading to site operations dominating the process. This is not a good description of some contemporary building. Components are made and even assembled off site. Components arrive on site as sub-assemblies or 'prefabricates' made to size with no expectation of site adjustment. Not only does this mean that the production process now becomes more related to manufacturing but also site practice becomes focused on jointing and fixing processes to allow for adjustment and fit. In yet another process, 'in-situ' forming of components in their final position can be used, as, for example, in the reinforced concrete frame. Off-site manufacture may be adopted for the temporary works and reinforcement, while the site operations are focused on forming.

While some highly prefabricated systems have developed, building is still likely to involve a mixture of traditional, prefabricated and in-situ processes. It is, therefore, important to be clear about the balance between manufacture and assembly, on and off site. This is then related to the availability of trained operatives, technical and managerial staff with the skills necessary to achieve the quality required within any constraints of time. These aspects of the resource base are discussed in more detail in Chapter 6.

Identifying stages

The process of production as a whole has been identified as taking processed materials, forming them into components and then assembling them to create the building. Production can be seen as a series of stages or, as they are known in production planning, activities, each defined by the creation of the next part of the building. Each one of these activities will take place in a sequence and be associated with a series of operations that suggest methods and sets of resources that will bring the construction to a natural point at which inspection may be made prior to starting the next stage.

The stages can be as large or as detailed as required depending on the stage of thinking about the production process. Even at the design stage it may be necessary to think through some parts of the building in some detail, perhaps because of the innovative nature of the design. Clearly, when the production process itself is being planned then monthly and even daily programmes will be required.

The reasons for breaking the construction down into activities are to gain a sense of the methods and resources that might be employed, which will determine the cost, time and safety aspects, as well as judging the ability to ensure performance in, for instance, working within tolerances. It is difficult to lay down rules as to how big these stages should be for it does depend on how novel the solution being suggested is.

In well-established solutions with well-known details, whole elements such as foundations, walls and roofs might be sufficient. If the solution is new (or at least unknown or even forgotten) then care must be taken to think through often relatively small activities to be clear that the suggestion is buildable.

When these divisions become very fine they become the domain of the individual operative. They become part of the craft or trade and can be sensibly left to the individual undertaking the operations. There may at first seem little need to understand production at this level. However, as detailing changes it may become difficult to achieve to an appropriate standard with current practice. An example of this has been suggested as the reason for some of the early failures of mastic when first introduced into construction. The specification called for surfaces to be 'clean'. Craft practice at the time gave meaning to the word clean that was not appropriate. The methods used to complete the operation of 'clean surfaces' involved products and tools unable to achieve the standards required for performance of the mastic. Without the operative being made aware of the important aspects of the specification or detail to achieve performance, common practice may be used in an inappropriate way which may jeopardise the ultimate performance. An appreciation of craft or trade practice is necessary to question the buildability of a suggested solution.

One other type of activity that may be considered in some detail right down to the operational level is one that is repeated many times in the assembly of the whole building. Small savings in these activities can bring high rewards in both cost and time without jeopardising performance if they are thought through at the time the solution is being evaluated. This type of analysis is important in highly prefabricated systems.

More generally, the depth of analysis required is associated with the risks involved. Risks are normally associated with safety, technical and financial aspects of a project. These are normally at their greatest with novel construction or where site conditions are complex, such as working in or near buildings that have to remain in operation.

Safety is a prime reason why it is necessary to carry out a risk analysis. The danger to health and safety at any stage of construction can often be reduced by changes in materials and details. The reduction of health and safety risks should be a constant aim in the choice of a solution as well as in the analysis of the production operations themselves.

Technical risk is associated with the possibility that the building will, through inappropriate production methods, not perform when completed or will suffer a premature failure either during construction or in the early years of its occupation. This is most often associated with not being able to achieve the required specification of materials, quality of workmanship or dimensional accuracy to ensure performance either in that part of the building or in parts to be completed subsequently.

Although both safety and technical risks carry a financial penalty, some risks directly affect budget expectations. Financial risk of this type can arise in many ways, often not just associated with the direct costs being more than anticipated. An example of this is financial penalties that arise from not meeting occupation dates. Risks to meeting occupation dates can be reviewed by identifying activities that are on the critical path. This defines the stages that form part of a chain of activities that represent the minimum length of time of the project. Delays in these activities directly affect the date on which the building can be handed over to the client. Planning techniques such as critical path analysis can be used to identify critical activities. Knowing in some detail the methods to be employed on these activities will lower the risk of running over time and incurring financial penalties, without jeopardising the performance of the occupied building.

Analysis of operations, methods and resources

Once decisions concerning the level of analysis required have been taken and an appropriate

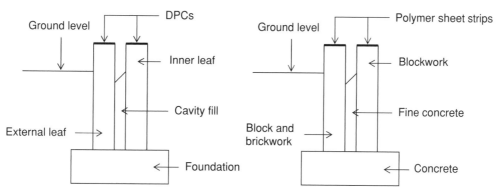

Figure 13.1 Strip foundations – parts and materials.

division of activities and a sequence identified, then the starting and finishing conditions of each activity can be visualised. It is then necessary to identify the operations required to complete that stage of the work. From an understanding of the operations, the methods and their associated resources can be chosen for the stage under consideration. However, the methods and resources may be influenced by those required for other stages, to give continuity. Repetition of operations and continuity of resources have an influence on productivity and the time–cost–quality relationship.

Figure 13.1 shows a cross section through the foundation to a house. A concrete strip footing supports the cavity masonry wall. While this is typically what would appear on the construc-

tion drawings there are other activities or stages in the construction that have to be undertaken to achieve the design, notably the removal of the topsoil and the excavation of the trench. These activities plus those associated with the construction that is shown on the drawings are illustrated in Figure 13.2. This illustration shows the sequence as the activities are shown on a production planning Gantt or bar chart.

It is now possible to identify the operations associated with each stage in the construction of the foundation. This requires the introduction of a new vocabulary to visualise and describe the proposed construction. Activities or stages are described by reference to the parts of the construction to be created: foundations, wall, damp-proof course (DPC). Figure 13.3 suggests

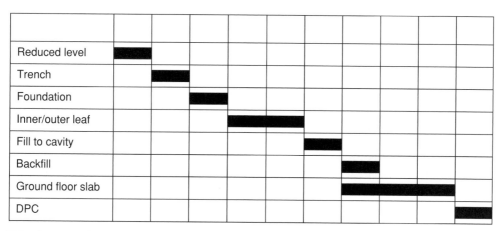

Figure 13.2 Stages and sequence of construction of strip foundations.

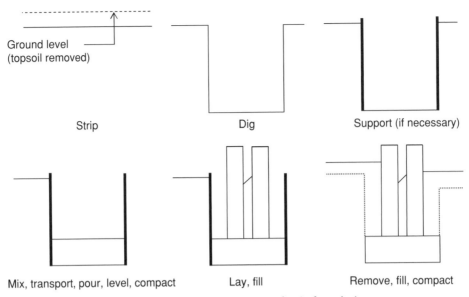

Figure 13.3 Operations associated with stages of construction of strip foundations.

that operations are actions and therefore need active words: e.g. dig, lay, pour, transport.

Figure 13.3 also illustrates how often the activities involve a number of operations. Trench would involve digging but may also require temporary works support. Temporary works do not normally appear on construction drawings. Only an understanding of production process allows the identification of the need for temporary structures and some ideas as to what methods might be appropriate and resources necessary to ensure safe and economic production. Forming the footings will involve the chain of integrated operations associated with concreting: mix, transport, pour, level and compact. All these have to be controlled to ensure that the quality of the concrete and the position of the concrete are within tolerances.

It is these operational actions that give clues to the alternative methods and the mix of resources that might be employed. For the excavation of the trench, hand methods might have to be employed in very confined spaces but are unlikely to be economic if access is available for machines to be used. If machines are used then the skills of the labour change to driver and banks man whose job it is to guide the machine driver and

ensure safe operations. Concreting is likely to be manual in the bottom of the relatively small trench, but options for the methods of transport will depend on the quantity, rate of delivery and site conditions. For small works, wheelbarrows might be appropriate, but again on open sites, machines for transporting the concrete will probably be more economic. Laying the DPC is likely to be part of normal bricklaying operations when the external walls are built.

Returning to Figure 13.2, the Gantt or bar chart is used to show not only sequence but also time. It is the method and the quantity of resources that determine the time an activity will take. This is clearly most important in production planning for the organisation of the appropriate resources in the right place at the right time in the right quantity. However, even at the design stage, clients are interested in time as well as budgets, so time often becomes a key aspect in the choice of a construction solution.

It is now possible to see the range of resources that might be used in manufacture and assembly of buildings. The direct resources involved will include the specified materials, the trained and skilled operatives required to complete the work and the equipment necessary to assist in the production process. Management and technical support will also be necessary with the requisite expertise and experience. Indirect resources associated with corporate organisation and capital will also be necessary but will not be considered in this analysis of the manufacturing and assembly in this text.

Resources may include the need for temporary works not shown as materials and components on the drawings of the completed building. It is often necessary to provide a range of temporary works and safety measures in order to complete the work. In some cases these temporary measures can form a significant part of the costs and time associated with the completion of the activity. This is the case in in-situ concrete construction. Temporary works have to be assembled and removed and this involves a series of operations requiring resources. In a detailed analysis of the production process they may be seen as activities in their own right.

The possible methods and mix of resources are primarily determined by the choice of construction materials and detailing. Production cost and productivity have a baseline set by the construction chosen for the design. Careful choice of a production method and good management can only create efficiency gains. It is at the stage where construction solutions are chosen that basic cost and productivity are established.

Materials and labour

Fundamental to nearly all construction activity is the notion of labour operating on materials to bring them closer to their final state and position in the completed building. It is a characteristic of industrialised societies that the materials will pass through a number of processes, worked by a number of different people. Each of these people will have different skills and expertise and may work in a variety of places, from winning the materials to final placing and finishing in the building. This division of work on the materials is then reflected in the number of organisations that undertake this work and the levels of technical and managerial expertise that need

to be applied. This in turn has implications for communication within the construction process, including those activities involved in the choice of construction.

The construction industry still relies on a number of traditional skills, such as carpentry and bricklaying, although there is now an increasing diversity of trades. These are normally more specialised, capable of a smaller range of operations, normally carried out within a well-defined stage of production associated with one part of the construction. In whatever way the skills are distributed among individuals, each operation will need to employ operatives with the required expertise and experience.

Health and safety

While labour remains at the heart of the operations for construction there is a major concern for the health and safety of the individual. Industrial operations carry with them specific risks. The analysis of the operations will reveal the hazards to which individuals will be exposed and the harm these hazards might cause them. These will identify the risks to individuals and therefore need to be part of any evaluation carried out in a proposed production process. The evaluation of the risks may lead to changes in the choice of the construction specification and details as well as to the need for safety measures being incorporated into operational procedures.

Risk extends beyond the individual carrying out the operations. Any work may endanger others in the vicinity of the operation, either on site or in its immediate environment. This includes not only the immediate danger of the accident but also exposure to both short-term and long-term health risks. Most of these risks are covered by regulations and operating a safe place of work is a requirement of the law. The law places responsibilities on various parties in the design and construction of buildings.

It is recognised that the most effective way to reduce risk is to eliminate by design. Once the design is established the next approach is to change people's behaviour, and then to devise safe working procedures during the production process. Personal protection is considered to be the least desirable option. Although personal protection is still a major contributor to reducing accidents and long-term health problems in construction, design should aim to reduce the individual's reliance on personal clothing and equipment for his/her ultimate safety.

The specification of construction materials and details can help to reduce this risk. Avoiding the specification of materials with known health risks from fibres, solvents, etc. can make a direct contribution to health and safety. Where the risk is very high, as with asbestos, substitutes were found. With materials such as cement, the risks are identifiable but lead to a situation where it is still specified, since alternatives are not easy to identify and protective measures, which are relatively simple, can be taken.

Detailing can help to determine where individuals have to be standing when completing fixings and the size and weight of parts they will have to handle. Loads to be lifted should be limited and working platforms should be stable and allow posture to be maintained. Working at height, particularly with loose parts, increases the danger of falling, requiring more measures against accidents. Risks exist when gaining access and egress from the working place, particularly if this includes carrying materials or components. Risks can be highest when operations are being carried out in close proximity, particularly if one is above the other. Other high-risk operations include working in confined spaces and lone working.

While this is not intended to be a comprehensive list of safety issues, it is meant to indicate that risk evaluation comes from an analysis of the operations dictated by the choice of construction specification and details. Methods chosen have to incorporate safety measures to create a safe working procedure. Real contributions to health and safety can be achieved in the choice of construction solutions.

Production equipment

Most building operations are still completed by an individual working on materials or

Table 13.1 Production equipment.

Temporary works	Plant and machinery
Structural support	Materials handling
● Partial erected construction	Excavation and earthworks
● Forming	Concreting operations
● Earthwork support	Manual operations
● Existing building support	Power plant
● Working platforms	
Site operations	
● Access and egress	
● Environmental enclosures	
● Security	
● Health and safety	
● Environmental protection	

components by hand, now probably assisted by powered tools. The worker will, however, almost certainly be supported by a range of production equipment. The two major categories of this equipment are temporary works, and plant and machinery. These have always been employed by constructors to make the work possible, but now also to make it safer, easier and to increase productivity. Many construction solutions are now only possible because machinery and support systems are available.

Typical examples of temporary works and plant and machinery are given in Table 13.1. This production equipment does not form part of the permanent construction. It is normally dismantled and removed once the permanent works can be used safely. It does not normally appear on any drawings or in any specification for the construction itself. It is, however, implied in the choice of construction and detailing. Changing the construction changes the need for temporary works and plant and machinery.

In most cases the economy of these production equipment systems lies in their ability to be used a number of times in different locations, often assisting in the construction of a number of different buildings. This means that any analysis of their useful life includes an analysis of the ease with which they can be dismantled and re-erected into similar, but often not identical forms. This process of removal and reforming becomes a major cost in the life of many items of equipment. This may be as straightforward as transporting an item of plant between sites or the complete dismantling and re-erection including retesting of falsework systems or a tower crane. This requirement of reuse has a profound effect on the approach that would be taken in their design, both in the standardisation of components of systems and, particularly, in the methods of jointing the components. Joints will be designed to be quick in both erection and dismantling. This can be seen in many temporary support systems such as scaffolding. It is these characteristics that determine the economy of a proposed construction specification or detail.

Production equipment has to be reliable. This is a relatively straightforward idea of breakdown for plant and machinery, but more generally the association of risk becomes linked to safety rather than loss of performance. In addition to aspects of durability, the use of production equipment means it is more prone to damage, misuse and unauthorised alteration. This potential for unsafe equipment creates the need for regular examination and even retesting at regular intervals.

While both temporary works and plant and equipment have many characteristics in common, they have their origins in two distinct functions associated with the production process. Temporary works can be identified as fulfilling a primary function in that without them labour by itself could not safely undertake the activities dictated by the construction solution chosen. Plant and machinery, however, have their origins in assisting in operations that were once carried out by hand. In this process of mechanisation, machines replace some labour (often associated with heavy, dirty or unsafe operations), although they normally generate new jobs, often of higher skill. Lifting equipment replaces large numbers of operatives on-site with drivers, banksmen, mechanics, etc. Much of what is now built could not be achieved by hand as solutions are adopted on the understanding that certain types of plant and machinery are available.

Increasingly, there are items of production equipment where the distinction between temporary works and plant and machinery is becoming blurred. The emergence of the MEWP (mobile elevated working platforms) is perhaps

the most striking example. While clearly appearing to be part of the process of mechanisation (a vehicle able to be powered to the required location), what has been mechanised (holding labour safely while they complete the operations) would have been previously provided as temporary works.

The distinction does, however, hold good for a large proportion of production equipment and can therefore be considered separately even if some equipment could be argued to be in either category.

Temporary works

Temporary works generally have either a structural support or a site operational role, including environmental enclosures and security, as indicated in Table 13.1. In all of these are elements of health and safety and this will influence the design, detailing and timing of the introduction and removal of these works.

First there are those systems that provide structural support until the permanent works can be used safely. Some forms of construction such as arches are constructed from components that only function as a whole when they are completed. This illustrates a fundamental question that has to be asked about any construction solution: 'Will the components be self-supporting as the assembly progresses?' If they are not there will be a need for some temporary works to support the partially erected structure. In the case of arches this is known as centring.

If the components are made from a material such as concrete that has no initial form it will need a mould. Concrete can be formed on site in the final position, where it is known as in situ. It can also be formed in a mould, often not even on the site, and then transported and fixed in its final position as a component; this is known as precast. For both in situ and precast a basic mould known as formwork will be required. If in situ the mould will have to be held in the position required in the completed building. This support is known as falsework. Since the materials from which the moulds are made (timber, metal or even glass fibre) are very different to the material to be moulded, different trades may be involved. Thinking through the mould and support requirement for this type of construction at the time of design, and detailing for simplification and repetitions, can have a significant effect on the cost of the final construction. This has been seen in the design of in-situ reinforced concrete slabs in the development from the beam and slab to flat slab designs, where the simplification of the formwork is the major source of cost saving.

Another major area of support that may be required is that of excavations. Because of the great variety of ground conditions and configurations of permanent works below ground this is a major area of consideration. Where excavations are deep and/or near existing structures, earthwork support often involves a full engineering design. The need for earthwork support has implications for both cost and construction time. The choice is often left to the constructor, where considerations of sequence and difficulties of constructing permanent works around temporary structures have to be weighed against the risk of ground movements during construction. However, the development of solutions such as diaphragm walling and contiguous piling, allowing the permanent works to be installed without traditional earthwork support methods, have made a major contribution to the time, cost and safety of the construction of basements particularly in city-centre developments.

When works above ground are adjacent to an existing building, or part of an existing building has to be retained, or alterations are being carried out, support offered by the original building may be removed. While the permanent works will have to reinstate this support, temporary support may be needed during the production process. This is normally known as shoring.

The temporary works identified above have all been to support part of the construction or the ground until the permanent works can safely take the loads themselves. The final group of temporary support systems with a structural role is working platforms. These do not support the construction itself but support the labour and materials associated with the operations as they are carried out. They are required until the

operations of assembly are complete. Working platforms are most often scaffolds and have to support the loads of the materials as well as the individuals themselves. Because of the direct use by individuals in order to carry out operations, health and safety becomes a major feature of the detailing of the scaffolding. It is important to consider not only the posture and ergonomics associated with carrying out the operations but also the possibilities of accidents.

Statistics show accidents from falling as a major source of injury and that most of these occur when work is being carried out at height, not only on temporary platforms but also from the permanent construction. Dangers from falling include both individuals themselves falling and materials or tools being used during the operations falling and injuring those below. This makes edge protection of particular significance. On temporary platforms such as scaffolds these have to be incorporated in the design as handrails and toe boards. If the permanent works form the working platform, specific temporary edge protection will be required.

Detailing can often dictate where the operative has to stand and what operations have to be completed on or near to the edge of the building. For example, fixings for cladding can sometimes be arranged to be on the top of the slab or on the face of the edge of the slab. Fixings can involve drilling or the use of cast in fixing channels. While these have implications for tolerances and achieving fit, they also determine the options associated with the operations to be carried out at the edge of the building.

Working platforms can be associated with mobile equipment (MEWP) more easily recognised as plant. These are of particular use when operations have to be carried out in a series of relatively isolated positions, at height, above a clear area, with an adequate surface for the equipment to move around safely.

Many temporary structures do not have an essentially structural role. They are more generally associated with the organisation of the overall site. The influence of the construction chosen is still very strong. Many systems will be associated with temporary access for the movement of materials, components and individuals to the place where the operations are to be carried out. Access (getting to) and egress (getting from) the place of work cover both horizontal and vertical movements. The type of construction will determine the range of loads and number of movements.

The analysis of these temporary works is at the heart of site organisation. Planning routes for the transfer of materials, areas of storage and methods of handling have to be considered with aspects of temporary access. The loads to be carried, the routes to be taken and physical size of access requirements determine these routes. Cranes carrying materials through the air need only landing areas, while hoists require support from either permanent works or, more likely, temporary scaffold structures. Vehicular access not only will need roads as well as hard standing for unloading, but also may need environmental control equipment to limit the spread of mud. Separating out pedestrian access from vehicular routes contributes to safety.

The link here to plant and machinery is very strong. The examples above of cranes and hoists for materials handling indicate how levels of mechanisation determine the need for temporary access and egress structures.

While the idea of individuals having to move both to and from the working area is fairly straightforward, it may at first be assumed that materials only have to be taken to the place of assembly. All operations create some level of waste. This may come from preparing materials, shaping components, cutting to fit or just from the packaging in which the components were delivered. Some estimates of the amount of this waste and the form it will take have to be taken into account when planning disposal.

Temporary construction may have to be provided for environmental enclosure. Nearly all sites will require temporary office and storage buildings, the extent of which will be proportional to the value of the works. It is possible on some sites to prepare materials or even complete sub-assemblies in areas on site, but not in the final position in the building. These operations may be covered, creating workshops or small manufacturing areas. It may even be beneficial to provide cover for the actual assembly process itself if adverse weather is expected, although the majority of assembly operations would be carried out under open site conditions.

Security works may be necessary at the boundary of the site not only to protect the works but also to limit access of the public to the potential dangers of site operations. Further levels of security may be required for storage compounds to protect materials and components from theft and vandalism. Some materials, such as bottled gases, pose particular safety hazards and these may carry a legal requirement for special secure storage conditions.

Although the temporary works so far identified will achieve the majority of health and safety provision, some operations may need specific provision. Particular areas of danger such as edges of excavations, overhead power cables or holes in slabs will also need protection for the unwary.

Many site operations may threaten the environment. This may include the soil, air, watercourses or the local wildlife balance. While this is best avoided by devising procedures that do not affect the environment, it may be necessary to provide specialist disposal and/or temporary screening or settling arrangements.

It has been suggested that some temporary works are necessary as support structures and some because of the needs of the overall site operations. In all cases they are influenced by the choice of construction. Analysing how a proposed solution has to be manufactured and assembled includes identifying the temporary works that the suggestion will require.

Plant and machinery

The distinction has already been made that while temporary works are fundamentally necessary to assemble buildings, plant and machinery have their origins in the mechanisation of manual operations. However, many buildings could not now be built without machines as the choice of construction is made in the knowledge that the plant and machinery exist. An understanding of the levels of mechanisation that are available has to be part of the analysis of the proposed solutions.

The most successful set of manual operations to be mechanised is materials handling. The size and weight of loads that can be lifted and carried have a direct influence on both the size of components and the potential size of building as a whole. The range and scope of materials handling plant are very wide. They vary not only in the size of loads they can carry but also in how versatile many have become, being able to move a variety of materials and components around the site and into position.

Perhaps the most obvious distinction is that between horizontal and vertical movement. Dumpers, trucks and lorries move horizontally, hoists and winches vertically. Cranes, forklifts and pumps and even helicopters can achieve both horizontal and vertical movement. All of this plant has an operational envelope defining the limits of movement in the horizontal and vertical planes that has to be matched to the site access and the form of construction chosen.

Excavation and earth moving is a specific site activity that has developed a range of plant for specialised materials handling. The action of digging involves additional forces over and above the action of picking and carrying the load. Machines range from the versatile slew machines with a range of buckets to specialist trenching machines. Pumps are used to mechanise the manual operation of 'bailing out' flooded excavations. There is a range of lorry and dumper transport options, again dependent on the size of operations.

Another site activity that involves a range of operations that have been mechanised is concreting. While most of the mechanisation is materials handling, other operations are associated with preparation and final assembly, for all but the smallest amounts of concrete mixing will be mechanised, ranging from small hand-fed mixers to fully automated batching plant. Moving the mixed concrete to the formwork can use a variety of materials handling plant depending on site conditions and the stage of construction. Once in the formwork the

compacting and even the spreading and levelling operations may be mechanised. One concreting activity that has become even more mechanised is the laying of large areas of concrete ground-supported floors. Not only is the plant highly specialised in placing, spreading, levelling and finishing, but also it incorporates some elements of automation.

Many operations remain essentially hand operations but are assisted by hand and even bench tools. Cutting, drilling, screwing and spraying are just some of the hand operations that are now commonly undertaken with power tools. Many of the smaller versions of these tools can be battery operated, but for larger, more powerful applications a power supply has to be available. On-site electricity and compressed air are the two main sources of power supply for small tools. This may lead to the use of generators or compressors as on-site power plants.

Most plant is either hand held (some simple machines such as winches are even hand operated) or controlled by a driver. There have been some attempts to automate the process with driverless plant using computer control. While this has found limited application on site, it is increasingly used in factory production. The use of computer-controlled technology in manufacture is leading to systems that link electronically produced design information with the manufacturing facility to automate the fabrication process.

Understanding the process and possibilities of mechanisation identifies the relationship of the construction solution to the production process. Developments and availability of plant and equipment have an effect on the possibilities for the development of the construction solution. In some cases the two may have to be developed together. In early civil engineering works the engineer would have designed machines for production as well as the civil engineering works themselves. Most current common construction is only possible because of the availability of site equipment and manufacturing techniques. Most common approaches to delivery of materials are only possible because of the availability of materials handling plant. This again demonstrates why production analysis becomes a necessary part of the evaluation of a suggested construction solution.

Knowledge and expertise for the analysis of the process

Carrying out this analysis of manufacturing and assembly requires knowledge and expertise gained from practice.

At the most fundamental level this knowledge is derived from craft and operative practice with modern methods, now including considerable use of plant and machinery from hand tools to high levels of automation. The basis of the understanding of the ways in which basic operations can be achieved comes from the know-how of those who work the materials and components and complete the assembly tasks. These are the people who know what might be achieved within the deviations demanded by the design and the time this might take. They will also have a view on temporary works and safe working procedures.

With modern specifications of materials this know-how has to be supported by technical knowledge. When new materials are introduced, the key aspects of the operations that control their properties, and therefore performance, may not be immediately apparent from previous know-how. This has to be worked out through the technical knowledge, although with time

this may become part of common practice. In some cases the technical support becomes part of common practice. An example of this is the determination of concrete strength to allow the decision to remove the temporary works.

This is then overlaid with knowledge and expertise gained by those who organise and manage the production process. They have the ability to take an overview of the whole production process. They have to see all the stages or activities that have to be carried out. Management will gain a view on sequence, temporary works, factory or site organisation and safe working practice. They have a perspective concerning overall productivity, cost and time. They then need to be informed by, and understand, the operative and technical knowledge to ensure it is integrated into the plans and programmes that they provide.

In the same way that the choice of the construction is guided by an overall design but has to be solved in the choice of materials and details, so production has to be guided by overall programmes but has to be solved in choice of methods and resources. These require a different knowledge base and approach to analysis, but it does not mean that these two creative acts should be carried out independently. Overall productivity is governed as much by the design as the efficiency of the production operations. Overall performance is governed as much by methods of production as the materials and details of construction.

Summary

(1) The analysis of the production aspects of manufacture and assembly is necessary as part of the evaluation of a suggested form of construction to ensure that the potential performance can be realised in the final working building.

(2) The analysis is based on visualising the building in pieces and at the various stages of assembly in its partially erected condition.

(3) How many stages it is necessary to divide the building process into will depend on the novelty of the construction and in particular the perceived technical risks. For productivity, repetitive operations may also be seen as worthy of a high level of analysis.

(4) Once stages are determined the analysis continues through an identification of the operations required to prepare and assemble the works and the choice of methods to be used which determine the resources necessary.

(5) The primary resources for construction work remain materials and labour supported by temporary works and appropriate levels of mechanisation.

(6) Because labour remains a key resource in production, health and safety is a major consideration in choosing methods and resources, both of which can be influenced by the choice of construction materials and details.

(7) Performance, quality, cost and time are determined by the sequence and methods; therefore the analysis of production must be an integral part of the choice of the construction materials and details.

(8) The knowledge to carry out an analysis comes primarily from the know-how developed from practice. However, this is increasingly being supplemented by both technical and operational expertise.

14 Cost

This chapter explores the issues concerning the analysis of the cost of a proposed construction solution. The ideas of cost and price chains and the need to identify the costs in a clear time frame within an economic and fiscal context demonstrate the necessity to be clear about whose cost it is and what the cost should include. There is no single definition of the cost of a solution. The most usual interpretation of the cost is the actual construction cost to the client. However, this may be misleading and design, running and even disposal cost may be a better basis for choice. This depends on to whom the cost applies and the value he/she puts on current and future expenditure.

Cost and value

It is clear that the cost of a solution is important if for no other reason than ultimately a solution that is outside the economic reach of the client will not be built. It is also true that clients are unlikely to want to spend more than they need to obtain what they want. Two solutions of similar performance are likely to be chosen on price. What clients want and what they can afford may also be a conflict that has to be resolved in the evaluation of a proposed solution. All of these need a clear basis for the analysis of costs.

What clients want, and the proportion of their money they are prepared to spend on it, is related to the worth or value they place on the service they get from the building. This sense of value or worth that clients are likely to place on different features of the building has to be established at the briefing stage, along with the more straightforward idea of the budget that will be available. Judgements on value may be made in different ways and these will influence the way in which cost should be analysed.

Perhaps the simplest notion of value comes from the provision of what might be termed as the basic performances, such as weatherproofing and fire performance. Here the value is in a level of performance that the client would probably suggest should be achieved at least cost. How-

ever, clients may be prepared to pay a great deal for the appearance of the building, for its value may be in the symbol or image it represents of the individual or the corporation. Other clients may not be prepared to spend their money on this. This is evident in the premium that clients are prepared to pay for tall buildings located in city centres. Other clients may value the quiet that can be achieved with acoustic measures (assuming location has been fixed) or the quality of light from additional money spent on lighting. These judgements on value are predominantly issues of design that need to be translated into solutions and here costs will be included which will have to be borne by the client. These design issues are often difficult to justify in cost terms as they can, in basic performance terms, be seen as useless. However, useless should not be confused with worthless, as it is in its value to the purchaser that the cost is justified.

There is another notion of value: one associated with paying more now for lower cost in the future. This is the argument for the value of thermal insulation, with its associated payback periods of lower heating bills. However, some clients may not be prepared to pay for insulation over and above the statutory requirements for they may wish to invest their current resources in other assets and be willing to fund future costs from future income.

Cost and price

The cost incurred in adopting a particular solution for the construction of a building is computed from the price paid for the various resources employed for the goods and services in order to complete the work necessary. Costs are created from a chain of prices charged which create the costs of each stage in the process. Hence the cost to the client for the construction of a wall is not the cost to the constructor. It is the price the client agrees to pay the constructor. The cost to the constructor is the price the constructor agrees to pay suppliers of goods and services needed to complete the works.

If, therefore, it is required to analyse the cost of a solution, the question that has to be asked first is to whom? The usual answer to this would be the client, for it is the client that has the resources to invest, and the expectation of a return on the investment.

The scope of much analysis of cost that is carried out for the client is limited to the cost for the construction itself; the price paid to the constructor for the final assembly. This is often represented by the 'tender' price. In the professional and industrial context there is considerable thought put into this concept of cost, for its commercial significance is high and hence data are available for its analysis. It is, therefore, tempting to use this as the basis on which a suggested solution for the construction should be evaluated for cost. This is, however, the client's cost, representing the price paid for the services from the constructor, and any evaluation will be based on historical tender data. This may not be the most appropriate basis on which to analyse the cost of a solution.

The two major limitations in applying this basis of analysis are, first, that the cost data are historical, making interpretation of the figures for new or novel solutions difficult, and, second, it does not take into account the design or running costs, let alone the disposal and redevelopment implications.

Cost and time frames

The construction cost forms only part of the outlay the client will have to make. Even staying within the same time frame of cost incurred after land acquisition and prior to occupation of the building, there are costs associated with designing the works. Generally, where construction solutions are well established, the design costs will be lower than for novel designs. If experience and the gathering of existing knowledge still prove inadequate to determine the performance with an appropriate level of confidence, then research and development costs may be incurred, although these are rare in construction generally.

The time frame of land acquisition to occupation also represents for the client a period of expenditure prior to gaining an income from the building. Designers and constructors will require payment in stages as the design and construction work proceed. The longer it takes to bring a building into occupation the higher this cost. If the time between initial investment in the land and occupation can be influenced by the selection of a construction it can be argued that this is a cost associated with that choice. Hence it could be seen as part of the analysis of its cost.

Different solutions have different costs after occupation of the building. Deterioration of materials suggested for alternative solutions will take place at different rates, requiring different renewal cycles to maintenance requirements. The reliability of components will vary, leading to different replacement costs. The detailing will determine the ease with which renewal and replacement can be achieved. This not only affects the direct cost of maintenance but also may carry a cost of disruption to the occupants while the work is carried out.

The overall building will have associated running costs directly affected by the construction choice. The example of the inclusion of insulation has already been mentioned and represents a part of the energy-use cost of a building. Another example is associated with the choice of surfaces, where the cleaning of a commercial building over its life may be many times that of the original construction costs. If cleaning can be achieved with less staff, albeit that the initial cost of the surface was more expensive, the overall cost may be less.

Finally, different solutions will have different disposal costs. This is not just the cost of demolition but also increasingly the cost of safe transport and deposit of waste from the demolition of the building. However, some materials that are waste to one project may have value to another. Recycling of architectural salvage has long had a commercial value, but increasingly building waste can also provide an income when the building is to be demolished.

There are cost implications associated with the land that has to be purchased by the client. The availability of certain types of construction solutions will bring land into economic use. This is particularly true in the case of foundation and geotechnical ground improvement processes that can make poor land economic or allow useable space below ground. This can also be the case if lightweight superstructure solutions can lower the cost of foundation solutions.

Increasingly the development of brown-field sites where previous development has left land contaminated is becoming more viable as technical solutions to cleaning or capping sites are developed. This is also true if projected to the end of the life of the building in order to consider the redevelopment costs. While it is unlikely that new development will be allowed to contaminate land in the way past users have, certain foundation solutions such as piles may be difficult to remove. This will make the value of the land less for redevelopment at the end of the life of the building.

The analysis of the cost of a solution has to be qualified by the question of not only whose costs are to be analysed but also over what time frame does this individual want the costs evaluated. It is unlikely that the full spectrum will be taken into account, as the client has to balance the use of current capital against future income. However, increasingly a fuller cost evaluation, often defined as life cycle costing, is being undertaken and is affecting the choice of the construction.

Cost and cost data

It has already been identified that while data concerning construction costs are readily available, they are predominantly based on historical prices for which constructors were willing to undertake the work once it is fully specified. The costs are based on measurements of parts of designed buildings and therefore rely on new work being similar to previous construction. There are reliable processes for estimating the cost of this measured work, but this is linked to construction costs of items measured in a standard way.

Extending the requirements of a cost analysis beyond that of construction costs into extended time frames has the difficulty of establishing reliable cost data. While design costs might be estimated, it may be difficult to extract costs for designing a specific part of the building. Feedback on the cost of running a building may be limited particularly for costs expected to arise some decades into the future. The lives of buildings are long and ownership often not continuous, so gathering these data is difficult. Even if these data were available there would still be the need to project this into the future where resource availability and economic conditions are difficult to predict.

A more fundamental basis of analysis

Costs arise from the deployment of resources. The price that has to be paid for resources depends on economic (and fiscal) conditions.

For example, resources to be deployed for manufacture and assembly can be identified from an analysis of the operations required and the methods that might be used to complete those operations. This is the analysis undertaken to identify the buildability of a solution identified in Chapter 13. It involves taking materials and components and operating on them with appropriate labour and equipment in a safe manner. This will give not only the quantity of resources required but also the time over which the labour and equipment will be employed. To these direct operational costs one must add any management and technical support and a proportion of the organisation costs, for these are also deploying resources.

In a market economy, price is primarily influenced by supply and demand; with the tendency to surplus, price falls, whereas with scarcity, price rises. In a competitive market the 'bid' price will be based on a judgment of what others might be asking for the equivalent goods or services. The fiscal conditions influencing costs are either direct taxation or inducements such as employment schemes or regional development grants.

The further into the future these predictions have to be made, the greater the chance of errors in the cost analysis. Running costs in the first few years can be identified with some certainty, but the long-term renewal and replacement costs are often some years or decades ahead. Alternative resources may become available (or existing resources become unavailable), which may make an analysis of the expected operations and methods inappropriate. Changes in demand and supply or in the presence of new alternatives increasing competition will distort price differentials. New taxation will change prices, as happened in the UK with the introduction of VAT on maintenance work as well as new build. At the end of the building's life, demolition may be made very expensive with the discovery of new hazards, as happened with asbestos, or made cheaper by the introduction of recycling methods, as with concrete.

All these mean that estimating costs in extended time frames can only be carried out with less confidence. The figures will be subject to possible errors and risks that need to be analysed before choices are made based on these extended time frames.

Building up the price

While this chapter is about evaluating cost, it has been established that costs are built up from a cost price chain where economic (and fiscal) factors influence price. It is therefore necessary to understand the generation of prices.

Price can be fixed in one of two basic ways: what the market will stand or the application of rational margins. In practice there are likely to be aspects of both in the setting of a price, particularly in bidding in a competitive situation. In both cases, knowledge of the expected costs is important, plus some estimate of how certain those costs will be. This allows some judgment on margins. The margin is the excess of price over cost that an organisation needs to survive and grow. The need for margins and the role of profit is beyond the scope of this text. Even if rational margins are applied to the costs, there is still the question of what the market will stand. If this is below costs then it may not be sensible to bid.

While, as has been indicated, a full analysis of pricing is beyond the scope of this text, there is one more aspect that could be usefully highlighted at this stage. The relationship between cost and price should also reflect the risk. This may represent the technical risk of non-performance of the completed works and the subsequent costs this will incur. It may represent a financial risk associated with delay and non-performance of the contract. For these and other risks it might be possible to identify insurance that is at least a known cost that can be included. If, however, the risk is to be carried, then some judgment on what price makes the risk worthwhile has to be made. Generally it should be possible to get a lower price if you are willing to take the risk. Who in the cost price chain takes what risk is dependent on the contract entered into. Designs based on manufacturers' components and calculations transfer risk to the manufacturer. Developments such as serial tendering and partnering have the effect of limiting or even sharing the risk through shared understandings and information between the contracting parties. This is likely to affect the construction solutions seeking lower costs for all parties involved.

Summary

(1) It is necessary to be clear what the client sees as value for money in establishing design and performance criteria, following which it is reasonable to seek economic solutions.

(2) Cost is not a single figure but depends on the answer to whose cost it is and what time frame the cost is to include. Cost could include construction, design, running and disposal costs, which can be further refined by looking at investment cost and fiscal advantages.

(3) In accounting for costs that will be incurred in the near future, price information may be available. Allowing for expenditure in the future requires knowledge of future technical developments and the economic and fiscal context and is therefore less easy to predict.

(4) Price may be based on market forces and/or rational margins but will also reflect the risk, technical and financial, that is to be taken in designing, constructing and operating the chosen construction solution.

15 Sustainability – Social Concern

This final chapter in Part 1 considers sustainability as a new social concern requiring a technical response. Specific technical knowledge concerning sustainable development is given in previous chapters where it needs to be embedded in the normal process of choice. Sustainability is here set in the context of social demands requiring solutions outside the current knowledge base and the need to understand the care required in making choices until this knowledge is established.

Two major forces

There are two major determinants of the construction solutions that we adopt. One is the law of nature and the other is the social context within which we operate. These two great forces condition the technology we develop and apply. This is the context in which we build. We strive to understand the natural resources that provide our materials and the forces of nature that act on the final construction. This is the knowledge we used in the analysis of the physical behaviour. Our knowledge in this area is now well established and, while it will improve, it is unlikely to change fundamentally. We then need to understand the cultural, economic and political conditions that exist and these will change with time. These shape the designs and determine the resources that are available for construction. They affect the commercial and ethical limits that govern the organisation of the industry. This influences the skills available and methods of operation that make solutions buildable and become the major determinant of the cost of construction. These changes may well require new technical knowledge with the need to return to our fundamental understanding of systems and systems behaviour.

Any extended discussion on the nature and impact of cultural beliefs and values and an ethical framework is beyond the scope of this text. It is most often so embedded in a society that it is covered in what would be commonly accepted to be a 'reasonable' solution. However, when

society is questioning its own values and ethical framework it is inevitable that these will be reflected in the questioning that has to be undertaken about what has to be 'reasonably' achieved and what is a 'reasonable' solution. This leads to changes that have technical consequences. Technology has to respond to shifts in values to be accepted by society in its own continual process of evolution and change.

As those involved in construction are usually part of the society they serve or are at least culturally linked to that society, most of these influences are inherent in their decision-making. This was not true when, in the second half of the twentieth century, the developed countries took their technology to the developing countries with the best of intentions, yet errors were made. It gave rise to the term 'appropriate technology' to remind those engaged with other cultures of the significance of the context of society with the technical solutions chosen. We should always strive to provide appropriate technology.

At the turn of the twentieth century the new social concern for our impact on the natural environment has emerged and is having a profound effect on buildings and the way we build. It is making us question what is appropriate technology for the twenty-first century, which implies changes beyond our current knowledge. This concern for the environment has been integrated into the idea of our ability to provide for the future without any compromise to the quality of life we have now or to limit the hopes of future generations. This overarching concern for the future has been encapsulated in the concept of sustainability.

Sustainability

All generations have worked to improve the quality of life for their children. As societies develop this has been dependent on an economic process, the creation of wealth, and then in its distribution that determines social equity, providing the rewards of economic development to individuals. Linked to political beliefs associated with values and ethics and the best ways to stimulate economic growth and social equity, this has determined the demand for buildings and provided the means to produce them. In recent years we have made such progress in economic and social terms that development is now at an unprecedented scale.

This has brought the two great forces, the natural environment and society, together. Development is now changing the environment to such an extent that we have to face the possibility that we will not have the resources to maintain our current rate of development or, even worse, may make the environment so hostile that we will not be able to provide technological solutions to maintain and improve the quality of life. It is now widely believed that the choices available and the quality of life for the next generation will be harmed if we do not bring together concerns for the natural environment as well as our concerns to maintain stable economic and social conditions.

This movement to bring concerns for the natural environment into our analysis and make it part of our process of choice has been termed sustainable development, normally just called sustainability. It sees the stewardship of the environment as equally important as the management of the economy and ensuring equity among the members of society. This accounting for the 'cost' of development in environmental, economic and equity terms is known as the 'triple bottom line' or the '3es'.

Sustainable development is not just about buildings, but buildings do have a significant impact on the environment as well as on the economy and within society and will therefore have to take the new concerns into account. In the same way as the choice of construction is mediated by economic and social conditions so it will need to take into account its impact on the environment.

Established influences and new concerns

Many of the economic and social influences seen as important in sustainability are well established and have been shaping our designs and the construction solution for many years. These would include the performance expectation of clients and the levels of infrastructure developed within a society for industrial activity. These are established but often changing and evolving. However, our concerns for the environment are new. These concerns are now evident in intellectual, political and social activity. These will need to be tested and given reality in the technological solutions to be put into practice.

This has generated the need to identify the important issues and look for knowledge and understanding sufficient to suggest 'safe' solutions. It may be that designers cannot analyse the behaviour, constructors fail to manage the critical aspects of quality or that the operatives have little experience or misinterpret the experience they have. The technological knowledge resides with all those involved. This text is prepared for the analysis and evaluation of choice and suggests that successful construction will require the existence of complementary knowledge and the know-how of all involved. It becomes part of the analysis of a solution to ensure that this complementary expertise exists before choice is made if performance is to be realised. This highlights the need for education and training and research and development to be a part of the culture of both society and the industry and professions that contribute to construction.

It is often difficult to differentiate between the established and truly new issues. It may be argued, for example, that the most recent changes in health and safety requirements to include design as well as production do not represent the fundamental change in our view of the value of life but reflect our improving economic standards, at least part of which should be spent on greater safety standards. The more fundamental change in values concerning health and safety took place with the public health reform, which also had a profound effect on construction solutions, changing not only detailing of buildings but also the site operational arrangements with provision for inspection and enforcement. It was part of the general movement towards equity to improve social conditions and ensure greater sharing of the new economic wealth being generated from the industrial revolution. Health and safety is now part of any analysis for buildability and the operation of a building. New requirements are absorbed and acted on within a broad base of knowledge and experience available to the designer, constructor and operator of the building.

Another example of a social concern that is having an influence on construction is 'inclusion' and in particular the need to take into account the needs of people with disabilities in our development of the built environment. This is essentially just an extension of our understanding of designing for people and is unlikely to have the far-reaching effect on construction that sustainability will have, but it will have an impact on the way we build. While it is not always easy to modify existing buildings, all the indications are that in new buildings we should be able to include everybody by modifying the solutions we already employ. Perhaps the simplest example is wheelchair access at external doors. The need for level thresholds will change the relationship of internal floor level and external ground levels unless ramps are used. By bringing the internal and external levels together, the potential for water to enter under storm conditions, particularly in flood areas, is increased. This will lead to new detailing around thresholds involving linear drainage gratings and possible flood defence measures designed into new buildings. There will be consequences for the levels of damp-proof courses (DPCs) and the detailing

of ventilations to suspended ground floors, but these are not new requirements, just modifications of well-understood construction.

The essence of an emerging issue such as concerns for the environment is a lack of a knowledge base and experience. There will almost certainly be no direct technological understanding of these issues among those engaged in the design and realisation of the building. This lack of understanding will range across individuals who hold the complementary knowledge, the total technology. These represent all those involved from researchers to craftsmen. Indeed it is difficult to identify who may have the most relevant knowledge to both think through and put into practice the first attempts at solutions.

An example of this was the drive for energy conservation following the oil crisis of the early 1970s. It was essentially an economically driven crisis that meant that society as a whole recognised, or at least did not question, the need to act. Social acceptance for the principle meant that political action was ensured which led to demands for a very rapid response from the technology. One response was to increase performance levels from insulation. In many ways the demand for this response was beyond the ability of the technology to develop and provide solutions that would maintain all of its performance expectations. While increased resistance to the passage of heat may have been achieved, the detailing and the site operations were not well enough understood to provide 'safe' solutions. With the increased insulation many cases of increased dampness were recorded, through either rain penetration or condensation, with new materials and detailing. There was a breakdown in perhaps the most fundamental requirements of a dry environment. It is helpful for sustainability, where reducing energy use and carbon dioxide pollution play a major part, that these early difficulties with insulation are now well understood and we now have many 'safe' solutions to increasing insulation values.

The concept of sustainable development, as defined at the beginning of the twenty-first century, introduces environmental impact into decisions on construction detailing and specification. The analysis brings into question the origin of the materials, their manufacture, assembly and disposal with its contribution to pollution and energy use as well as direct issues of resource depletion and changes in ecosystems. Sustainability requires the analysis of the operation of the building and its impact on the environment in pollution and energy terms over the life cycle of the building. It questions the source of energy, for if energy can be made available economically with little or no environmental impact, then it may be used as part of sustainable development. Further it demands this analysis be undertaken against global and regional as well as local impacts on the environment.

There becomes a need to act in conjunction with other agencies who are considering sustainability at other levels such as settlement development, transport, etc. Sustainability also looks at the impacts over extended periods of time with its underlying objective of leaving resources for following generations. It demands data and information that are currently incomplete, if not yet known, further adding to the difficulties of making informed decisions about construction solutions at this time.

Society will expect technology to respond and play its part and possibly be unforgiving of failure. However, much of this new analysis will involve knowledge of the process of production and assembly and of the materials themselves. In terms of this book the issues of sustainability have been introduced as part of the evaluation of the design, the physical behaviour and manufacture and assembly. As knowledge and experience grow, they will become a natural part of the choice of construction. Treating them as special considerations runs the risk of the errors made with insulation detailing following the oil crisis of the early 1970s.

Of the two great forces suggested at the beginning of this chapter it is the social forces that change. Sustainable development is the new concern of the early years of the twenty-first century. As we learn more about the issues and devise solutions that prove themselves in practice and provide precedent, the need for environmental concerns will become an established response in our designs and production procedures. Sustainability is just the latest in the history of the need for technology to respond to new social concerns. It is unlikely to be the last.

Summary

(1) There are two major forces that condition a construction solution: the laws of nature and the society in which construction is practised.

(2) At any time in a society there may be deep changes in the values and beliefs, which have to be taken into account in a technical solution for it to be seen as a reasonable option. At the beginning of the twenty-first century this includes embracing environmental concerns in our understanding of sustainable development.

(3) These more fundamental shifts in values often demand changes for which knowledge is uncertain, making the analysis of physical behaviour of the construction even more significant to ensure 'safe' solutions.

Part 2
Choice

16 Applying the Framework to Housing

This opening chapter for the second part of the book affirms the process of choice determined by the earlier chapters and establishes the approach taken in the final chapters as a case study based on making a choice of construction for a house in the UK at the beginning of the twenty-first century.

The need for an integrated approach

There is a danger in supposing that because the previous chapters suggest that construction has to be evaluated through a number of areas of analysis, then they can be considered one by one and can be carried out in a given order to reach a choice. This view is not correct and it is hoped that this text has not given this impression. The first few chapters of the book are designed to give a holistic feel to the process of choice before the detail of analysis of construction is introduced.

In reality each project has its own balance of priorities set by the client and its own context that determine the appropriate approach. The emphasis is on evaluation as a holistic process, not satisfactorily complete until all the areas of analysis show acceptable predictions. While each analysis is largely independent, it is always being undertaken in a growing understanding of the outcomes of the evaluation as a whole. Early proposals need to be provisional until the outcomes of each analysis have become clear. This leads to the need to have techniques to establish approximate solutions for each analysis in order to see if an overall solution might be possible as well as detailed design information to make the final choice.

This text is based on a framework, not a sequence. The process has been thought of as dynamic, open and iterative, moving towards a solution but without prescribing the route. The choice has to be made within a framework that indicates the range and scope of issues that have to be taken into account, a framework that can be used to establish the possible complementary, possibly competing factors that will determine the success of the construction in technical and social terms.

It is difficult to reproduce this process in the pages of a book. It is, by its very nature, one that has to be experienced to fully understand how what sounds like a chaotic process is controlled and rational, yet creative choices are made in practice.

This text has set out to explain a basis for the technical and rational aspects of the process and maintains that it is only through this understanding that the creative and innovative can be put into practice.

In order to continue to apply this approach, the chapters that follow are written from the perspective of a case study. The case is the choice of construction for houses for the UK at the beginning of the twenty-first century. This chapter introduces the context in which these choices are being made, the nature of the environments, physical and social, and identifies the aspects of design, appearance and the resource base that will condition the choices to be made.

The basis of the case study

The study is focused on the house, a single-occupation dwelling with its own ground floor and roof, and therefore does not include flats, although the house could be detached, semi-detached or terraced. This description could include specialist accommodation such as sheltered housing, so the study will be further limited to 'family' occupation. Given the time and place specified for the case study, the notion of a family, its members and their relationship is changing. Because the house is still the aspiration for many of these families, this text assumes that these changes will not significantly affect the construction of what these families are seeking, although there may be changes in the performance requirements with new patterns of use within the house.

While this description of a house will allow a broad understanding of the needs of the user, it says nothing about the resource base or industrial structures to both design and build houses. This case study will try to embrace the decision making for the speculative, volume house builder and for the individual, client-based house designer. This wide spectrum will allow consideration of variations in choice based on the economic power of both client and constructor.

The two ideas used so far to define this case study, family and industry (demand and supply), have set much of the scope of choice. The depth of understanding that is necessary, of each of these ideas, should not be underestimated to appreciate why certain construction solutions will be chosen.

This was the thrust of some of the earlier chapters of this book where it was suggested that it is necessary to understand something of the physical and social environments within which choices are made. This will include appreciation of house design and the resource base available to the house-building industry.

Physical and social environment

The case study is set in an advanced industrial society on a highly populated island in a temperate climate. The social values include the distribution of private wealth with equity and respect for all members of society, with a growing awareness of environmental impact. This gives a high level of demand for housing, with limited land and high prices on the open market but a need for social housing.

These few facts start to create a picture of the context in which the choices will be made. Many of the details will emerge in the chapters that follow, but even this short description can lead to some more observations. The advanced industrial society means that we have a great deal of legislation and regulation. We can take advantage of factory production and transport of component and prefabricated sections of buildings from almost anywhere in the world. We are, however, having a high impact on the environment. The need for housing on our highly populated island means that we are increasingly building on previously developed land, brown-field sites, and have a problem in dealing with our waste. Our concerns for equity and social justice, with a fair standard of living for all, set relatively high performance levels for construction. The temperate climate defines our weathering problems and indicates that water penetration and winter heating are major concerns for house construction. There are now concerns that the levels of human activity we wish to sustain are causing environmental changes that threaten our ability to maintain this progress. Environmental concerns must take their place with economic and social considerations to ensure sustainable development.

There is limited value in this text in exploring these ideas in detail, as this is not the purpose of this book. The more significant aspects have been introduced in the previous chapters and the reader should become increasing aware of the importance of these factors as the following chapters develop the basis of choice for the construction of houses.

The resource base

With high levels of house-building activity the supply of materials and factory-produced components is good, but there is a shortage of site labour and the traditional trade skills in particular. To meet demand considerable attention has been paid to specifying components and prefabrication including modular and even volumetric systems to limit site operations (although these will not be discussed in detail in the chapters that follow). Machinery is readily available for materials handling. The introduction of cordless hand tools has influenced fixing and jointing processes, often now the focus of the site assembly processes. Money for housing is readily available where risk is low although the resources for social housing are more limited and highly regulated.

Design and appearance

The basic design and appearance of the house is well established. The pattern of living is sufficiently common for the major rooms of living room, bedroom, kitchen and bathroom to be easily identifiable, with some circulation space including hallways and staircases. While there is increasing variety in the number of rooms in a house and some creative internal layouts can be observed, the size of the rooms has generally been reducing, particularly in the volume housing market where the whole building footprint and plot size is reduced. This has led in some cases to habitable space extending into the roof.

Externally the brick house is still a dominant form. Rendered finishes might be used, although they still maintain the sense of masonry construction. Roofs are pitched with tiles or slates to give the characteristic shape to either bungalows or two- and three-storey houses. There is then a variety of window forms including bays and oriels often used to gain variety in housing estate design. It is not possible to generalise on the designs for the individual client, where expression in materials such as glass and timber may be featured along with a willingness to include aspects associated with sustainable development.

Internally, finishes are plain but need to accommodate a variety of building services. Energy saving has featured in the regulations over recent years and has had a large influence on detailing, and this is likely to continue with increased environmental concerns. In the single dwelling, fire resistance is only nominal within the dwelling, but walls between dwellings (party walls) and the roof finishes that are continuous between dwellings are subject to more stringent regulation. New regulations on soundproofing and air infiltration with the possibility of the need for performance testing and the introduction of 'robust' detailing are also changing the construction of and specification for houses.

Environmental concerns are affecting the balance between passive and active technologies as well as the choice of the components, materials and detailing itself.

Common form

The conditions established for the case study have given rise to the common form introduced in Chapter 3. This will be developed in greater detail in the chapters that follow. With the overall form established it is possible to take an elemental approach to floors, roofs, walls, foundations and services. Each element contributes a major group of functions and is easily recognised by the basic construction solutions. However, the question arises as to what order they should be considered in. It has already been established that no one element can be fully selected without some reference to the probable form of the others. In this case the elements that can be discussed knowing least detail of the other elements are the floors and the roof. These transfer loads to the walls so it is necessary to discuss these before walls can be fully investigated. It is then possible to complete the structure by looking at foundations where the loads from a house are relatively low and the site conditions are unlikely to dominate the choices for the superstructure. Finishes will be discussed with each element to cover all the analysis of the passive construction. This leaves the active services, although aspects of them will have inevitably been introduced by their need to be integrated with the other elements.

Summary

(1) The framework identifies the areas to be considered but does not suggest the sequence in which the areas of analysis should be undertaken. The nature of each project will determine the key choices, with no single choice being confirmed until all areas have been considered, at least in outline.

(2) The case study approach to be used in the next five chapters concerns the choice of construction for a house in the UK to be built at the beginning of the twenty-first century.

(3) The physical and social environment in which this case study is being considered is represented by the needs of the family in an advanced industrial society, on a highly populated island, in a temperate climate, in a society that strives for social equity, and with a growing environmental awareness.

(4) The advanced industrial society gives high labour costs and shortages but provides the infrastructure for factory production and prefabrication as an alternative to traditional craft-based construction.

(5) Given the design expectations of the user housing, a well-defined format for the house has emerged for the speculative and social housing market, although much of the detailing within this format will be evolving and must be confirmed for each project.

17 Floors

This chapter introduces both upper and ground floors for the house construction that might be adopted in the case study outlined in the previous chapter. It follows the analysis of the choice of construction detail and specification, suggesting alternatives and giving some typical details to support and illustrate the analysis.

Upper floors

It is sensible to consider upper floors and ground floors separately. They have a different set of functions to fulfil and the structural support options for upper floors where the spans are defined by room layout do not apply to ground floors.

Upper floors are structures spanning over the rooms below and providing a stable clean surface for the rooms on the upper stories. Being internal they have no weathering function and are unlikely to be involved in resisting the passage of heat or in providing light or ventilation. While they will not contribute to these passive environmental functions, many of the services distribution systems may need to be integrated into the space or zone occupied by the floor.

Within a dwelling, security is unlikely to be a performance requirement for the floor. However, with changing patterns of living and the expectations of individual members of the family, privacy, particularly sound transmission, may be an issue. This may not be satisfied by just a structure with the addition of the finishes. Where floors are between dwellings in flats or apartments they will have greater fire and sound requirements which will have a significant effect on the choice of construction.

General forms for upper floors

There are two general forms that might provide an economic solution to this set of performance requirements: joisted floors and slab floors. Joisted floors are based on small beams (joists) set a limited distance apart with an additional component, the floor board, to form the continuous surface required to support a finish for the room above. The floor would normally also have to provide support for a ceiling, probably of plasterboard, for the room below. A slab is a solid (typically concrete) continuous structure that, while normally spanning only in one direction like the joists (one-way spanning), can be designed to transfer the loads on all four edges (two-way spanning).

In housing, floor spans would not normally be above 5 metres and that makes the timber-joisted structure economic, although greater spans are possible in timber if the shape of the joist is changed, as discussed later in this chapter. For longer spans, but particularly for the increased performance demands of sound and fire required between dwellings, the concrete slab might prove a better choice. Slabs in concrete can be formed in-situ but are more likely to be precast, either planks or a form of slab known as a beam and block construction. The beam and block floor has been adapted for suspended ground floors (discussed later in this chapter).

Timber joisted upper floors

For the upper floors of the house in this case study, a timber joist floor will be chosen. The basic arrangement and variables are shown in Figure 17.1. The floor structure illustrated shows

Figure 17.1   Timber joisted floor – arrangement and variables.

a traditional arrangement of rectangular section timber joists at regular centres with a boarding to provide the floor surface. Two of the basic variables have therefore been established: the material and shape of the joist. While the shape is unambiguous and only needs defining by size, the quality of the material needs further specification. Just specifying the joist as 'timber' is not sufficient for a structural (or indeed any other) application. Some way of ensuring its strength in order to determine a safe and economic size is necessary.

Timber is a variable material and one way it varies is in its strength. Even the basic distinction of hard and soft wood is not a good guide to strength. The species of the tree will give some indications, but variation still exists. A safe design would have to assume that the weakest timber produced by that species might be delivered to site and used in the construction. Timbers for structural purposes are stress graded where they are given a designation that guarantees the strength within a limited band, so the size can

be chosen with some confidence of the strength of timber that will be used in the construction. In the UK the two most common strength grades for structural timbers for this application are SC3 and SC4, with 4 being the stronger and therefore potentially smaller sections, although the timber itself is likely to be more expensive.

The size of the joist now depends on the centres and the ratio of the breadth to the depth of rectangular section. Larger centres mean fewer joists; larger sections and thicker boards as the centres determine the span of the boarding. In practice, centres between 400 and 600 mm are economic. This leaves the ratio of breadth to depth of the joist. The depth has a large influence on bending strength and limiting deflections. Breadth influences stability associated with lateral torsion buckling and to some extent how 'live' or dynamic the floor feels. There is little option but to invest in depth for bending strength and deflection, but there is an option to help improve stability and the overall stiffness of the floor rather than just making the joist thicker.

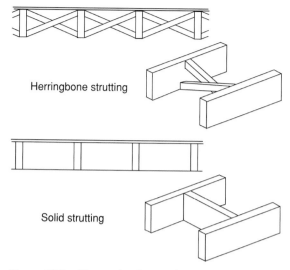

Herringbone strutting

Solid strutting

Figure 17.2   Alternative forms of strutting.

Figure 17.2 illustrates two forms of strutting, but both serve the same purpose. They stabilise the compressive zone from the bending forces in the top of one joist using the stable tensile zone at the bottom of the adjacent joists. This allows slimmer joists than would otherwise be necessary for stability and stiffness. The amount of strutting will depend on the breadth/depth ratio of the timbers selected for the joists. Joists are most likely to be selected from sections between around 38×100 mm and 75×225 mm, the larger sections becoming expensive and possibly hard to obtain. With these joists strutting is unlikely to be required up to spans of 2500 mm, with one line of strutting being sufficient at the centre of the span up to 4500 mm and then two rows at the third points above this span. It is unlikely that the timber joisted floor will be used much above this span so more than two rows of strutting are unlikely to be required.

As has already been identified, the selection of the joist is not unrelated to the choice of boarding. The board span is related to the centres of the joist, but first the material for the boarding must be chosen. Traditionally timber was the choice, but now it is more likely to be a composite sheet material such as chipboard. In this case defining strength quality is established by specifying flooring grade chipboard, but some

issues of durability might have to be addressed if the chipboard is likely to become damp. The integrity of a composite material such as chipboard is influenced by the binder or glue used and some may be susceptible to damp. Where damp resistance is required in areas such as bathrooms and kitchens it must be part of the specification. Whatever the durability specification the thickness of the board with joist at 400 or 450 mm centres, in a flooring grade chipboard, will have to be 18 mm thick and for 600 mm centres 22 mm thick. Alternative board materials such as plywood or oriented strand board (OSB) may be used.

Joints and fixings

Now the materials and the basic size and shape of the components have been chosen, and their spatial relationship defined, there is now a need to be clear about the joints and fixings to be employed. These will occur between:

- The sheets of the boarding
- The boarding and the joists
- The joists and the strutting
- The joists and the support

The two long edges of the chipboard sheets are tongue and grooved and glued together as these edges will not be fully supported, as they will be laid across the joists. The other joint, on the ends of the boards, should rest on a joist to give full support. The boards are then nailed or screwed to a specified pattern that, with the glued jointing, not only holds the boards in place but also eliminates squeaking when walking on the floor. If nails are used they should be at least 3 mm ring shank nails for improved grip in the board with a length 2.5 times the thickness of the board. To allow for movement, boards should not be cut hard up against the walls, but have a gap of 12 mm to the wall for an average room size and 3 mm under the skirting to allow for unimpeded expansion. Strutting is also either nailed or screwed to the joists.

The junction between floor and wall requires the transfer of load from the joists to the wall. The joint has to provide sufficient bearing so

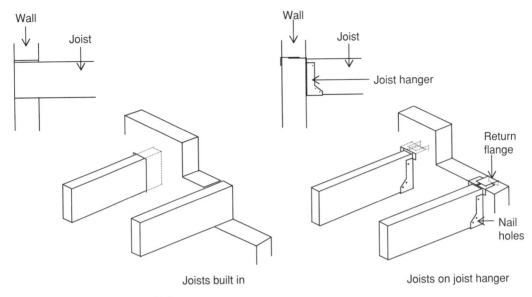

Figure 17.3 Alternative forms of joist support.

as not to cause excessive pressure (stress) at the point of transfer, which might lead to a crushing failure in either the end of the joist or the walling material. There is no particular requirement for the joint's structural action, but a simple pin joint would make the behaviour of the joist and wall more straightforward. The two options given in Figure 17.3 show, first, the simple building in of the end of the joist to give a full 100 mm bearing onto the wall and, second, the use of a joist hanger made of galvanised steel that can be specified with a return flange to provide additional stability to the wall. Both these simple, pragmatic connections provide a pin-jointed action so that the beam remains, in structural terms, simply supported.

Holes and openings

As with most elements of construction there is a requirement for openings, holes and voids to be formed. Small holes to pass services vertically through the floor can be accommodated between joists and through the boarding. There are few limits as to where or how big these holes can be so long as the edge of any cut (as opposed to drilled) board is supported on noggins, cross

timbers fixed between the joists, at the edge of larger voids. For staircases or chimneys the opening is greater than the space between two joists so the structure will require trimming. The arrangement for trimming and the use of hangers to make the joints are shown in Figure 17.4.

To run pipes and wires for the services systems horizontally, the space between the joists is available. However, if services have to run across the joists some portion of the joist will have to be removed for the pipe or wire to pass through. Care is necessary not to affect the structural integrity of the joist. The joist is a beam and the way it is connected to the wall makes it simply supported, so the greatest bending stresses are in the top and bottom of the beam at the centre of the span and the associated shear is greatest along the neutral axis towards the supports (see Chapter 11). Removal of timber at these points is not advisable. Some guidance as to where holes and notches can be made is given in Figure 17.5. As bending stress reduces towards the supports, notches can be removed from the top (or bottom but not both). Towards the centre of the span reduced shear makes holes on the neutral axis possible. Across the centre of the

Figure 17.4 Arrangements for trimming openings.

Repeat allowance
on other half of joist

Notches on top in a zone
between 0.1 and 0.2 of the
span maximum depth of 0.15
depth of joist

Holes on centre line (neutral axis)
in a zone between 0.25 and 0.4 of
the span maximum diameter of
0.25 depth of joist holes to be kept
apart at least 3x hole diameter

Notching and drilling in the
same joist should be at least
100 mm apart

Figure 17.5 Allowable areas for removal of timber from joist for services.

span the design will have assumed a full section to resist the bending moments and, while the material near the neutral axis is stressed less than that at the top and bottom, it is contributing to the overall bending capacity. Any reduction in this area may compromise the load carrying capacity of the joist.

Return to original suggestion

This analysis began with the cross-sectional shape of the joist being taken as a rectangle. If solid rectangular timber sections are selected then the decisions on detailing and specification are all very logical, but, like all initial decisions, it limits the subsequent possibilities: in this case particularly limiting the economic span to around 5 metres. If greater spans were required without changing the basic material then returning to the shape of the cross section might prove useful.

A couple of other useful cross-sectional shapes for beams might be adopted: the 'I' section and the box section. Both provide solid material at the top and bottom where bending stresses are greatest, give depth to limit deflection, while removing solid material to provide a more light-weight structure with less material. They do, however, have less material along the neutral axis and therefore have limited shear resistance, but on relatively lightly loaded longer spans this is unlikely to be the critical failure. Overall stability and dynamic behaviour is also likely to be less easy to achieve than with a solid timber section so this will also have to be revisited.

Both of these alternative cross sections have been developed, but the 'I' section has proved to be more commercial. These are fabricated sections based on a composite timber board web and timber sections, often laminated to form the flanges, as shown in Figure 17.6. The use of composite and laminated timber products reduces movements such as shrinkage, bowing, twisting and splitting associated with solid timber sections, but as the basic material has not changed, the majority of detailing and fixings can remain the same. Boards can be fixed as before and joist hangers are options for the connection to the wall. If these sections are used as trimmer joists

or doubled up as trimming joists or partitions support, then blocking at the web will be necessary where the loads are transferred and at intervals to stiffen the pair of joists. With the 'I' section the web contributes less in bending stress but has to take all the shear stress. This means that notches for services are not sensible at any point along the flanges, but there are more options for holes particularly at the centre of the span and even near the supports if additional shear stiffening is provided.

The 'I' section beams are manufactured under factory conditions and would be ordered to length for each project. On site they are less stable than the solid section timber joists when laid out before the boards are fixed, so temporary strutting is necessary for the boarding operation.

Life cycle considerations

Reflecting on the proposal so far, the issue of behaviour over time has only been considered for the possibility of dampness in the chipboard. Some more thought needs to be given to the main components so far suggested. Floor structures like this would be expected to last for the life of the building without any major maintenance. Having used the space for services distributions systems, however, these may need access for renewal or change to the pipes and wires. It may be better to consider the layout of the services and provide access hatches in the boarding, as the continuous nature of the glued joints makes subsequent partial access difficult.

Laminated
timber flanges

Depth from around
250–400 mm

Oriented strand
board (OSB) web

Available for approximately
5- to 8-metre spans on
domestic loading

Figure 17.6   'I' section timber joists.

Timber will deteriorate if the moisture content rises above 20% or conditions exist for a major infestation. This is unlikely for the bulk of the timber. However, if the timbers are built into the wall for support there is exposed end grain into the cavity. This will probably be protected by the insulation, keeping it both warm and dry, but if detailing suggests that this is not the case it may be better to apply preservative to the end grain on site, essential if the whole timber is specified with a preservative. If the timber is supported on hangers the timber is likely to be cut to length on site, so end grain protection may still be necessary if the whole joist is impregnated with a preservative. The hanger will be galvanised and should, therefore, not be vulnerable as long as the galvanising is not damaged in site handling and installation.

Dismantling at the end of the life cycle should not be difficult and the joists may be re-used if the timber is still in good condition or possibly sent for conversion to new chipboard or some other recycling options. Given the gluing and nailing of the chipboard sheets, removal may be difficult so disposal of the chipboard may not include re-use, always a possibility with old timber floorboards. Hangers might be recycled or cleaned for re-use at the end of the life of the house depending on their condition.

Finishes and other elements

While some idea of a suitable ceiling finish using plasterboard was in the initial suggestion it is necessary to return to the question of finishes to see if the proposed structure can provide the background for both floor and ceiling finishes.

It is necessary for the background for the finishes to provide:

- Sufficient stable support
- Suitable fixing opportunities
- Compatible flatness tolerances
- Matching for future movement
- Avoiding deterioration mechanisms

For the floor the most likely finish is carpet. This needs full support and will follow the tolerances of the background. The boarding can provide these, but the surface may be rather hard (lack of resilience) so an underlay for the carpet will have to be specified. It is possible to use an adhesive to stick carpet down and this would be possible on the chipboard, but a better solution is provided by an edge fixing nailed to the chipboard and this will overcome any relative movements. There should be no physical or chemical interaction between the carpet underlay and the chipboard.

For the ceiling the underside of the joists does not provide continuous support but can provide good fixings at reasonable centres. The initial suggestion of the plasterboard will be compatible with the solution suggested so long as it is thick enough not to sag between joists. Plasterboard provides a good surface to paint or apply an additional finish. With the board fixed firmly to the joists, initial movements may develop cracks at the junction of the ceiling and the wall where a coving may be used to cover and hide any cracking. All this can be achieved with a 12.5 mm board screwed or nailed to the joists.

While the integration of services has been considered above in the section on holes and openings, it is necessary to see if any other elements of construction might interact with the floor. In Chapter 19 the need for restraint for the external wall will be discussed and the floor can provide this. The hangers shown in Figure 17.3 can offer restraint where the joists transfer loads to the walls (note the need to nail the end of the joist to the hanger to ensure the joist provides restraint to the wall). Where the joists run parallel to the wall, straps can be provided connecting the inner skin of the wall to the first three joists, effectively holding the wall to the whole floor given sufficient fixing of the boarding to all the other joists. This detail, given in Figure 17.7, should be provided at a maximum of 2000 mm centres.

Internal walls that are not taking loads from the floor itself have to be built up to the underside of the ceiling. If these walls are made of blockwork they are stable without any fixings at the top of the wall as long as the ceiling is not

Strap housed and fixed into top of joist

Packing between last joist and wall

Internal skin of cavity wall

Wall tie

Joist

Noggings between joists

External skin of cavity wall

Figure 17.7 Strapping of external wall to floor.

too high, which is unlikely in a modern house. However, if the partitions are lightweight stud-work they will need fixing to the underside of the floor structure to provide stability to the head of the partition. If the partition is running at right angles to the joists then fixings are available at each joist. If a partition is running parallel with the joists then, unless it happens to be on the line of a joist, no fixings are available. In this situation it is necessary to provide noggins in the floor, fixed between the joists at between 400 and 600 mm centres, specifically to support the head of the partition. These have to be shown on the working drawing for the floor, as, like the strapping for the external walls, they have to be fixed by the carpenters as part of the floor construction.

If lightweight partitions are built from the floor then they can be fixed to the boarding. Block partitions at right angles to the joists will spread the load across the floor, but where a heavy partition is required in the direction of the joists it may be necessary to provide a double joist immediately below the partitions.

Sound transmission loss

Changes in patterns of occupation show that bedrooms are increasingly being used during the daytime when other rooms below are also in use. This is changing the performance expectations of the floor particularly for airborne sound reduction. The floor construction so far described may have been satisfactory in the past when bedrooms were used only at night but would fail to meet this new performance expectation. Performance levels for transmission loss (sound reduction index) of R_w of 40 dB are now expected from floors within dwellings. It is possible to achieve this performance by introducing a 100 mm thick sound-absorbent material in the cavity between the joists resting on the ceiling board. This, along with the other aspects of specification, is shown in Figure 17.8.

75 x 225 mm grade SC3 timber joist at 600 mm centres

22 mm flooring grade chipboard

12 mm plasterboard

100 mm mineral wool

Figure 17.8 Typical domestic timber joisted floor construction for around a 4.5 metre span.

The inclusion of sound-insulating material will influence the space available for any pipes or cables being distributed in the floor cavity. It may now be difficult to use holes along the neutral axis so services distribution will have to be across notches only allowable towards the supports. If electrical cables are laid under the quilting its thermal insulating properties may cause over-heating and will have to be taken into account in the sizing of the wires (see Chapter 21).

Ground floors

The choice of the construction for upper floors within a single dwelling discussed above was concerned primarily with establishing a structure capable of providing support for the imposed loads, stability for adjoining walls and a background for finishes to provide clean and stable surfaces with an adequate level of sound reduction. It is unlikely to have any other major function that determines the specification and detailing. This is not the case with the ground floor. Figure 17.9 shows diagrammatically the conditions to be expected below the floor and the conditions required above the floor. This indicates a number of different functions, including damp-proofing and resistance to the passage of heat. The conditions below the floor are potentially more aggressive to the deterioration of materials and the risk of movement is greater. All these have to be taken into account in the choice of any proposals for the construction of the ground floor.

While the upper floor has to span over the rooms below this is not the case for the ground floor. The ground conditions may provide suffi-cient strength and be stable enough to place the floor directly on the ground. This surface of stable ground has to be reasonably close to the desired ground floor level. If achieving stable ground requires excessive excavation or fill, a suspended floor may be a more economic option.

If a suspended floor is deemed to be the economic choice, the construction for the structure is likely to be similar to upper floors, but it is now possible to build walls to support the floor at any desired position below the structure. More choice is available to select the span, giv-ing an opportunity for economy. However, the additional functions and the more aggressive environment will make more demands on the detailing and the selection of materials.

Suspended ground floors

In traditional detailing, the floor would be a timber joisted structure with similar details to the upper floor construction. In a ground floor it is possible to gain some economy in the size of the joist by providing intermediate support on sleeper or honeycomb walls built on their own foundation with a damp-proof course (DPC) under a timber wall plate laid to receive the joists on top of the wall. This is shown in Figure 17.10. On top of the joists a board material would be provided much the same as an upper floor. Under-floor ventilation and the sealing of the ground surface under the floor are important aspects of detailing for the timber solution. There is potential for damp air to build up under the floor, increasing the moisture content of the timber and leaving it susceptible to fungal attack. This can be avoided by a good flow of fresh air below the floor and a ground-covering layer below the floor. Although the ventilation is required to ensure the durability of the structure and not directly to provide clean air for the occupants, the mechanisms of ventilation dis-cussed in Chapter 10 are being exploited in this detailing. They rely on natural ventilation gain-ing cross flow driven by the wind. This requires openings in the external walls on all sides of

Figure 17.9   Ground floor – conditions.

Figure 17.10 Suspended timber ground floor.

the building with external walls and an un-interrupted path for the air from one side to the other. Air bricks in the external skin, a duct across the cavity and then the holes in the sleeper walls provide this path for ventilation. The space between the ground cover layer and the under-side of the joist has to be at least 150 mm and at least 75 mm to the underside of the wall plates on the sleeper walls. Air bricks have to provide an opening at least 1500 mm^2 per metre run of wall and be designed to prevent an infestation of rodents, as the space below a floor would be very attractive to animal life. These details are also shown in Figure 17.10.

An alternative solution, a concrete beam and block floor based on joist-like beams with the surface formed with concrete blocks laid between them, has been developed. The general form, shown in Figure 17.11, has been taken from upper floor systems developed for commercial building where spans and loadings are greater. However, a system of components suitable for the smaller scale of the house floor has proved to be economic.

Figure 17.11 Beam and block ground floor – components.

The durability risk is reduced with the use of concrete instead of timber and therefore a ground cover layer is not necessary. However, under-floor ventilation detailing similar to the timber solution is still required. New recommendations are emerging, concerning areas where flooding is likely, that suggest that a means of inspecting and clearing out the under-floor space should be considered.

Where there is a risk of gas below the floor it may also be necessary to provide a gas barrier on top of the floor structure protected with a screed. In these circumstances it will also be necessary to protect the wall cavities with gas-proof DPC cavity trays. The nature of these membranes and other provision for gas control are discussed in the section on ground-supported slab floors under 'Environmental control and life cycle' below.

Resisting the passage of heat

The analysis so far of both the timber and concrete has shown that they have the potential to be economic structures, can control damp and the specification can be drawn up to ensure durability. If the walls are on sound foundations, the movement risk is now very low. However, they are not good at resisting the passage of heat, and the concrete beam and block will not have the required accuracy to apply a finish directly to the structure.

It will be necessary to introduce thermal insulation to resist the passage of heat. There are a range of materials that can be considered as good insulators, but their choice will not just rest on the best insulation value. It is, for instance, necessary to understand the form that insulating materials take: they can be loose fills, quilt, bats or boards. These vary in the stiffness of the layer they form, its compressibility and susceptibility to damage. These will determine what support is needed, the required fixings and aspects of the installation process.

Another consideration in the choice of material arises from environmental concerns. While one of the perceived needs for high insulation values is to reduce energy use and contribute to sustainable development, the materials chosen will have an impact on the environment during manufacture and right up to their delivery to site and installation procedures. The analysis of sustainable construction must account for energy at all stages of the life cycle as well as pollution, waste and any direct effects on habitat and eco-systems throughout the life of the material. The rate of resource depletion and the consideration of renewable or recycled resources as well as the ultimate disposal as waste or the potential to recycle have to be investigated and understood. This may lead to the choice of a material with lower insulation potential that can be provided as a thicker layer to achieve the same insulating effect.

Yet another consideration when analysing proposals to insulate is the position of the insulating layer. While the overall resistance to the passage of heat (the U value) is not influenced by the position, the temperature gradient across the construction is. This is explored in Chapter 10 where the need to consider thermal capacity, condensation risk and cold bridging is also explained.

Practical considerations would lead to an initial suggestion that the insulation layer should be positioned on top of the floor as indicated in Figure 17.12. It can be installed when the house is weather-tight on a stable, cleaned surface. However, a condensation risk now exists across the floor structure. Condensation on the underside of the structure will be limited by any cross ventilation, but detailing a vapour check layer on top of the insulation will be necessary to minimise the risk of condensation below the insulation.

The insulation on top of the floor will be subject to the imposed loads of the occupants. It will have to have some strength and a board material may be the only option unless a cross battening solution is considered, but this will introduce cold bridges, requiring thicker insulation to achieve the same U value. Even with a board material, impact or point loading will damage the insulation and the surface will be neither stable nor clean and will not provide the fixings for the finish such as carpet. It will be necessary to 'float' (unfixed) a floor boarding material over the insulation on a vapour check layer. An example of this could be the flooring grade chipboard with tongue and groove, glued joints on all four sides and a suitable movement gap at the edge under the skirting as the upper floor. The finish can now be fixed in the same way as on the upper floors. These details are also shown in Figure 17.12.

Figure 17.12 Beam and block ground floor – typical details.

If the insulation is to be placed below the structure it will have to be suspended in some way, and if placed between the joists in the timber solution it will have to be thicker as the joists will form cold bridges, requiring more insulation to gain the overall U value. With insulation in this position the condensation risk is now reduced and other forms of insulation such as quilts or loose fills can be considered, but opportunities for edge detailing to reduce cold bridging may be limited if the structure is transferring its load to the external walls.

Having identified the position and type of material it is necessary to specify a thickness. This will depend on the target U value and the conductivity of the insulating material, but the transmittance of ground floors is complicated by the way the heat is lost through the ground. Towards the perimeter of the building the heat will be lost back to the surface, while at the centre this effect is considerably reduced. The overall heat loss is dependent on the ratio of the perimeter to the area of the floor. The calculations for this are complex, but tabulated information is available to determine the required thickness.

Finishes

While this completes the analysis for the timber floor with all aspects evaluated, it leaves one question about the concrete beam and block solution: the required accuracy to apply a finish directly to the structure. If the insulation has been placed on top of the structure and a floor boarding floated on top of the insulation then this requirement is satisfied. An alternative to a board finish is a 50 mm thick sand and cement screed. If a screed is used the carpet may have to be fixed with an adhesive, as nailing into a screed is not satisfactory.

Ground-supported slab floors

As the name of this solution suggests, the structural behaviour of this floor is not in bending but compression as imposed loads are transmitted directly to the ground through the slab. This simple view of all compressive forces assumes that the conditions for three other potential failure modes are not present: (1) that heavy impact or point loads will not 'punch' through the slab; this is a shear failure but is very unlikely in a house and can be discounted; (2) that if the chosen material is subject to shrinkage, then the tensile forces this creates can be resisted without cracking the slab. The most likely material for the slab is in-situ concrete and this does shrink on curing. However, the size of the floor for a house is very unlikely to mean that the tensile forces will build up to create cracks of sufficient size to be a problem. This is, however, the case for large commercial floors where these tensile forces determine much of the detailing; and (3) that if the compressibility of the ground is not equal under the entire slab, this uneven support may induce a bending action as the stiff slab bridges across soft areas of ground. This condition may exist on some sites and does have to be checked. This lack of strength or variable compressibility may come from natural variations in the ground, but these are likely to be small over the area of a house if the ground is of a reasonable bearing capacity. If the soil is weak, or the floor is being built over ground filled at some earlier date or it is on shrinkable clay then the possibility of this mechanism becomes much higher and needs to be taken into account. Another cause of this uneven long-term support can arise where fill (hardcore) is required to make up levels during construction. The process involves compacting in layers and if this is not controlled, then settlement of this fill will increase the chance of the slab cracking. If any deep fill is required to make up levels then a suspended floor may be a more economic solution.

As has been indicated, the structural slab will be constructed in concrete. If the ground is stable there is no need for this to be reinforced, as the compressive strength of the concrete will be sufficient to resist the forces involved. If the ground is suspected of being weak or variable under the slab then reinforcement can be introduced to ensure loads are distributed and any differential settlements do not distress the slab.

Reinforced or not, if the slab of concrete is built directly on the ground it will risk some cracking due to the relative stiffness of each material. For this reason the slab needs to be laid on an intermediate layer, stronger and stiffer than the ground but less rigid than the concrete. This material, often referred to as hardcore, is a fill of graded inert loose material such as quarry waste or crushed recycled construction waste compacted without any cementation binder. For a house on reasonable ground this will only need to be 100 mm thick and can be compacted in one layer. On top of this a floor slab between 100 and 150 mm thick will be sufficient for house loadings.

Environmental control and life cycle

Having established a satisfactory structure it is necessary to return to the information identified in Figure 17.9 to start the analysis of the environmental control elements of the floor. The construction of a concrete slab on hardcore will not stop damp rising through the construction to the floor surface and will provide little thermal insulation. It is not easy to change the specification of the hardcore to gain much advantage in its thermal properties, and while lightweight concrete for the slab would have sufficient strength and improved insulation it would prove more expensive.

Damp resistance and insulation plus provision for a background for the finish will all require additional layers of materials with the right properties and of sufficient thickness to achieve the performance. Figure 17.13 shows one of the alternatives for these layers. It shows damp-proofing and the insulation between the hardcore and the slab and a sand and cement screed on top of the slab to provide the surface for the floor finish. An alternative is to place the insulation on top of the slab with a floated chipboard floor, as shown in Figure 17.12, over the beam and block floor. The damp-proof membrane (DPM) can then be either under the insulation or under the slab with a sand layer (blinding) over the hardcore to protect the DPM from damage. Whatever the position of the DPM it is most likely to be specified as a 1200-gauge polythene sheet.

The polythene membrane can also provide a barrier for radon gas protection so long as the correct grade is specified. If a gas barrier for methane and/or carbon dioxide is required then a multi-layered sheet including an aluminium foil and a layer to improve tear resistance has to be specified. These membranes will also act as the DPM. For gas barriers, joints should be sealed with double-sided tape and the cavity should be sealed with a gas-proof DPC tray. Particular care to seal the membrane at service entry points is also required and is discussed below. The gas membrane represents a low-permeability barrier that may be used in conjunction with a

Figure 17.13 Ground-supported slab floor – typical details.

high-permeability granular layer below the slab (in place of the hardcore) from which the gas can be removed in a controlled manner with passive ventilation.

The basis of choice of materials for insulation is similar to that explained for the suspended ground floor. Insulation placed on the slab will need to have a floating floor board or screed to provide the clean flat surface for the finish. Insulating the edge of the slab against the external wall to eliminate cold bridge is important. This heat loss path through the edge is almost directly to the outside air, representing a high proportion of the potential heat loss through the floor. As with the suspended floor the thickness of insulation required will depend on the ratio of the area of the floor to the perimeter of floor against external walls.

While the conditions below ground are the same as below a suspended floor, the materials that are being used are less susceptible to deterioration as the aggressive generators of decay for the hardcore and concrete are not automatically present. However, some care has to be taken in checking for harmful conditions. Given that the temperature of the soil under the house is unlikely to go below freezing, then even if the DPM is above the slab, frost attack is not a problem. The presence of chemicals is probably the greatest risk. The most aggressive to be found naturally in the soil are sulphates which react with the cement and disrupt the concrete unless a sulphate-resisting cement has been

specified or the permeability of the concrete is very low. This is not likely to be the case with the relatively low grade of structural concrete necessary for this type of slab. With increased use of brown-field sites some chemicals that are aggressive to hardcore or concrete may be present from previous occupations. For housing development, soil remediation will have to be undertaken to make the site safe for habitation and therefore many chemicals will be removed by pre-treatment or removal of the soil.

Another source of chemical attack may be brought in with the materials themselves. In talking about the fill layer it was indicated that the material should be inert. One of the major problems with 'hardcore' from traditional demolition waste is that other materials such as plaster remain. With the advent of procedures to separate demolition materials for recycling and then some processing, including crushing, this recycled material is now much safer and can be specified with the confidence of using a controlled material. The other source of chemical attack comes from within the concrete itself, with the use of reactive aggregates in the concrete leading to alkyl silica reactions.

Services and other elements

Having considered the basic form of the floor, be it suspended or ground supported, it is necessary to think about the way other elements of the construction will interact with the chosen solution. Internal partitions that do not have their own foundation will be built from the floor. It may be necessary to strengthen the floor where partitions will be built. Thickening of the slab in a ground-supported floor or double joists in a suspended floor might be required for heavier partitions. The relationship with the external wall will depend on the role of the wall in taking load from the floor. If the wall is used to take the load from a suspended solution then the details will be similar to the upper floors. For the ground support solution the walls may contain the floor during the construction process but will have little interaction during the operational life of the building.

There will be interaction with the incoming and outgoing services: water, electricity, gas, communications and drainage. Each of these will come from different depths below ground and will have a minimum radius for the bend between horizontal in the ground to vertical in the building. They may also have to be maintained or replaced during the life of the building. These requirements are fairly easy to accommodate in the suspended floor but have to be detailed very carefully in the solid ground-supported slab. The provision of ducts cast into the slab for incoming services will need to be included and outgoing drainage will have to be incorporated in the slab before it is cast. While this is relatively easy, special care has to be taken in keeping the continuity of the DPM around these services penetrations. Particular care has to be taken if the membrane is specified to be gas proof, with the use of preformed top-hat units to seal around service pipes.

Site influences on choice of ground floors

Analyses of behaviour under load, environmental control and durability show that there are a number of solutions for a ground floor. The choice of the most economic is likely to be determined by the site. Many house builders have standard designs for the house but will have more than one option for the ground floor, which will be selected only after the site conditions are known.

The site conditions include not only the ground and the slope of the land but also the access, as this will affect the production operations. Much of the site investigation will be undertaken to determine the choice of foundations, and this will be introduced in Chapter 20. It is necessary to understand the production operations in order to undertake this aspect of the analysis before the choice is made.

Production operations for ground floors

Both the suspended and the ground-supported floor are constructed after the foundations have been completed, with the walls constructed to damp-proof course (DPC) level. This includes any internal walls with their own foundation including sleeper walls for the suspended floor. The initial excavation will have removed the topsoil and reduced the level within the building to the underside of the construction for any ground-covering construction for the suspended floor or the hardcore fill layer for the ground-supported slab. All the operations to construct the floor therefore take place from the ground and within the external walls of the house. However, materials have to be prepared, stored and moved around the site.

The full construction of the floor will not take place in one continuous sequence. The basic structure will be completed before the external walls are built up to first floor level. The finishing work will be undertaken after the building is weather-tight. These later operations are not considered in the broad analysis provided below.

Materials movement and storage varies with each solution. Timber is relatively light to handle but requires specific storage to ensure it is not overstressed and that its moisture content does not increase. Chipboard must also be kept dry and insulating boards, while they may not be directly affected by the wet, will be easily damaged. The concrete components of the beam and block floor are more robust but can still be overstressed if not stored the right way up and are heavier to transport and position. The fill material for the ground support slab will need to be spread, levelled and compacted in layers and then blinded, for if the surface is not fine enough it might damage the DPM. This should allow the DPM to be laid easily, but care must be taken in sealing the DPM at joints and around the services and then at the edges to ensure continuity with the DPC. The concrete for the in-situ slab has to be mixed, poured, spread and levelled and then cured.

Each of these operations requires different skills. The timber joist requires carpenters while the beam and block requires perhaps a more limited set of skills associated with ground workers, although they may be installed by individuals with more traditional skills such as bricklaying to give continuity of work between walls and floors. Ground workers may construct the concrete slab or perhaps a dedicated concreting gang may be employed, although their skills are more associated with structural frame concrete operations and are unlikely to be used for more straightforward house construction.

All these operations will require temporary access for both machines and operatives. If possible the two should be kept separate to minimise the risk of accident. The type of machinery will vary. Heavy transport such as ready-mix concrete lorries will need to get close to the building so they can discharge directly to the slab of a ground-supported floor. Most materials will be offloaded from the delivery transport and then moved on site by smaller transport, most likely based on the forklift. Suitable running surface will need to be provided for each of these transport and handling options.

As the work will take place at ground level, scaffolds will not be required so the main hazards will be associated with handling the materials. Only the wet concrete would be considered as a hazardous material requiring protective equipment for the operatives over and above standard personal protection. Collision with materials handling plant has been identified and is perhaps the greatest risk as the harm caused is likely to be great. Manual handling risks are perhaps greatest with the beam and block floor where the beams weigh around 35 kg per metre run, making not only back injuries but also trapped fingers probable hazards.

The final choice

It is now possible to make the choice based on both performance and production criteria to match to the site conditions. However, one other consideration may influence choice. This is the scale of operations. For the single house the builder often had a particular mix of skills and may want to employ a more versatile machine to be used for many different operations. On a site with a large number of houses the opportunity for standardising and more economic scheduling of labour and machines will affect the best mix of resources.

Summary

(1) Upper floors span over the rooms below, providing a clean, stable surface and make a contribution to privacy in reducing sound transmission between rooms above and below. They will be used to route services but will have few other functions within the single dwelling being considered in this case study.

(2) The two general forms for upper floors, joisted and slab, can be considered, but currently the timber joisted solution is likely to be the economic choice given the performance requirements within a dwelling.

(3) The joist is traditionally a rectangular section in a structural grade timber with strutting to give stability to the floor. Boarding is most likely to be a sheet material, typically flooring grade chipboard, specified as moisture resistant in potentially damp areas.

(4) The boarding has to be joined together to make a continuous surface and then fixed securely to the joists. The connection of the joist to the supporting wall can be achieved by building the joist into the wall or being held in a hanger.

(5) Openings in joisted upper floors that are greater than the distance between the joists have to be trimmed. Holes to pass services horizontally through the joists are limited in both size and position to maintain the structural integrity of the joist.

(6) For longer spans, 'I' section joists constructed from laminated flanges and a board web can be used.

(7) The joist and boarded floor provides a surface for the finish for the rooms above and provides support to fix a board material, normally plasterboard, to the underside to form a ceiling. The floor is now sufficiently stable to provide a diaphragm to support the external walls. Internal partitions below may require head fixings and partitions above may require a double joist if running parallel with the joists.

(8) Ground floors have different performance requirements, mainly from damp control and resisting the passage of heat, and are in a more aggressive environment with respect to deterioration of the material.

(9) These ground floors can be suspended joisted solutions or ground support slabs. If suspended solutions are adopted then concrete beam and block offer an alternative to the traditional timber form. Ventilation for the void below the floor is an important detail for suspended construction.

(10) Insulating materials to resist the passage of heat can be specified in many forms that vary in the stiffness and compressibility of the layer and this is important if the insulation has to take the imposed loads from the room above. The position and thickness also have to take into account condensation risk and cold bridging.

(11) The ground support slab constructed with in-situ concrete, possibly reinforced, will be laid on a layer of compacted fill or hardcore. It will require a damp-proof membrane (DPM) normally placed under the slab. Insulation can be introduced in a number of positions with detailing at the edge important to eliminate the potential cold bridge.

(12) In addition to performance requirements, site conditions and production considerations will also influence the choice of construction.

18 Roofs

This chapter focuses on the pitched roof, although flat roofs are introduced. The general forms of both waterproof coverings and roof structure are discussed, but the chapter provides a full analysis of concrete interlocking tiles on a trussed rafter structure as the most likely solution for the case study.

Pitch and span

The choice of construction for a roof is dominated by the pitch and span. The span (distance between supports) influences the structural solution, the rainwater collection and disposal required and the movement to be accommodated, particularly in the waterproof coverings. The pitch (angle of the roof surface) has a major influence on the specification of the waterproof covering, will dictate the form of the structure and have a significant effect on the appearance of the building.

Roofs are normally considered as 'flat' or 'pitched'. The use of the term flat roof is not a very good description as the angled surface of most pitched roofs is flat and even a flat roof has to have a 'fall', i.e. is not level, in order to ensure that rainwater is effectively shed from the roof surface. However, it is a useful distinction in so far as 'flat' indicates that the fall is so small that it is not perceived as pitched, so the structure can be treated as a floor and the waterproofing covering has to be continuous with all the joints sealed.

For houses designed for the UK the level of rainfall and the strength of the wind are likely to make the pitched roof the dominant form as this can employ economic waterproof coverings that are more reliable, making it a long-lived and generally a more robust solution. It does, however, make the requirements of the structure more complex than the simple joisted (or slab) solution that can be adopted for a flat roof. While both the design of the house and the availability of an economic covering will dictate the actual pitch, the widespread adoption of the pitched roof is part of the characteristic appearance of the UK house.

Other functions

So far, two parts of the roof have been identified: the waterproof coverings (with rainwater disposal systems) and the structure. These two are linked by the need of the structure to support the chosen covering so that it maintains its waterproof function. In most designs there will be a need to provide a ceiling for the room below that will also be supported by the structure. The roof that is just waterproof and has a ceiling is very unlikely to provide sufficient resistance to the passage of heat, so consideration will have to be given to the thermal performance.

While the roof space has in the past been used to house service tanks and has by some householders been used for storage, it is increasingly likely that designs will call for living space to be incorporated in at least part of the roof. This will not only influence the structure and the detailing of the ceiling and insulation but also introduce the need for light, through either windows that extend above the roof line or roof lights installed in the plane of the roof surface. These will call for holes to be formed in both structure and covering. Even if light is not a requirement, holes may be required for chimneys and other service pipes, but these are likely to be smaller than for windows and roof lights. These aspects will not be covered in detail in this text.

When sustainable development is considered, the roof, particularly when south facing, can provide a surface to collect renewable solar energy, in the provision of an area for both solar panels that warm water to heat domestic hot water and arrays of photovoltaic cells to generate electricity (see Chapter 21). These units have to be fixed to the roof above the coverings and will require services (pipes and cables) to pass through the roof.

In providing physical security and privacy the specification and detailing for the other functions will probably give sufficient performance unless local conditions prevail: for example, sound transmission if the building is under the flight path near to an airport. However, the risk associated with fire has to be considered in all terraced and semi-detached designs that require separating walls (see Chapter 19) between the dwellings. Fire can spread along the surface of the roof coverings over the separating wall. The separating wall can be extended above the roof line, but this creates the need for weathering details with initial expense and a possible maintenance requirement. If the covering passes over the separating wall then it must be chosen to resist the spread of flame and detailed to ensure the fire is stopped over the top of the wall. This will also apply to the detailing at the top of the wall to ensure that sound transmission loss is maintained across the separating wall.

Pitched roofs

Before any discussion of pitched roofs can take place, it is necessary to understand the vocabulary that is used to describe the configurations of pitched roofs that are required to cover a range of plan arrangements. The basic configurations and terminology use are given in Figure 18.1.

Eaves are edges at which water is collected for disposal, normally in gutters connected to downpipes to take the water to underground disposal systems. Verges represent edges where water is not expected to be collected and are most usually associated with gable ends. Ridge, hip and valley are all formed where the pitched surfaces meet. While the ridge and hip shed water away down each of the roof surfaces, the valley effectively collects water. It has to be constructed to ensure the water on the roof is taken to the eaves. If the tiles themselves cannot be 'swept' around the valley to maintain the weatherproofing of the covering, a gutter has to be formed to collect and direct the water to the gutters and downpipes at the eaves. Abutments are the meeting between roof surfaces and walls, so the joint between them has to be sealed so that the water runs to the eaves and not into the roof space.

These junctions may need detailing with special tiles and components in other materials, for example flashings and soakers in lead at

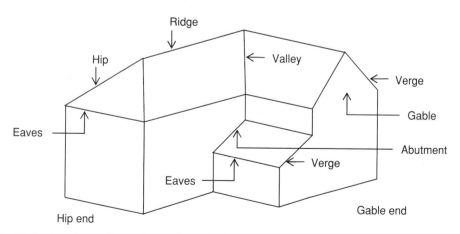

Figure 18.1   Pitched roof – configuration and terminology.

Table 18.1   Alternative roof coverings.

Generic types	Common forms
Continuous covering	Built-up felts
	Single-ply polymers
	Sheet metals
	Asphalt
Large sheet	Corrugated metals
	Corrugated plastics
	Corrugated cement compounds
Small tiles, slates or shingles	Clay or concrete tiles
	Natural or synthetic slates
	Timber shingles

abutments. Gutters formed at valleys have traditionally been in timber and lined in lead, although preformed sections in plastic materials can be used, fixed directly to the roof structure.

Coverings

There are three generic types of roof coverings, each of which has been developed into a number of common forms for which details for specific buildings can be developed. These are given in Table 18.1, which indicates the following forms:

- Continuous coverings
- Large sheet coverings
- Slates and tile coverings

The continuous coverings are all made from impermeable materials where all but asphalt, an in-situ hot laid process, can be formed into thin, flexible sheets to be laid on the roof making them light in weight. These thin lightweight sheets will require continuous support and fixing against wind uplift and, if used on pitch roofs, against sliding. Fixing can be achieved either by adhering the material to the support base or by mechanical fixing. The joints and edges have to be sealed to be weather-tight, either by adhesion as with bitumen for the built-up felts or by dry folding of seams on the metals. This action of folding the materials, forming welts and drips, relies on bringing the materials into close proximity on the folds to overcome both gravity and capillarity which would draw the water into the unsealed joint. Asphalt is spread onto the con-

tinuous support in layers to a total thickness of around 18 mm, making this covering significantly heavier than the thin sheet alternatives. The joints become sealed in the cooling process although joints are also staggered between each layer.

This combination of impermeable materials and sealed joints means that these coverings will work on flat as well as pitched roofs, although asphalt will slump with time and is therefore rarely specified for pitched roofs. These coverings have commonly been applied to flat roofs for small domestic buildings (built-up felts) and to commercial buildings (asphalt, metals and single-ply polymers) as well as to pitched roofs on more prestigious buildings (copper and other metal sheets) including some housing.

Large sheet coverings are also made of impermeable materials and therefore in relatively thin sections. These sheets are, however, stiffer by the nature of the material, its thickness and the shape into which they are formed. The sheet is given strength to span between supports by shaping the material into corrugations. Although corrugations were originally created based on the curve, giving the characteristic shape of 'corrugated iron', many of the metals now have folding based on the rectangle. Whatever the shape of the corrugations the sheets need to be fixed to the supports against wind uplift and sliding. These sheets are normally laid with both side and head joints lapped, so that they need not be sealed so long as the lap is large enough and the pitch sufficient to maintain adequate drainage runoff. The corrugations are used to assist weathering on the side lap as they are lapped on the crest of the corrugation. On lower pitches, head and even side laps will need to be sealed with strip mastic as the sheets are laid. Modern sheet materials can be colour coated as a direct finish.

Tiles and slates are made from materials of variable permeability. Natural and artificial slates can be considered to be impermeable and can therefore be used in thin sections. The material is rigid but cannot be shaped and therefore the strength of the material alone will limit the span between supports. Because the slate is flat, the side laps have to be 'butt' joints although 'lap'

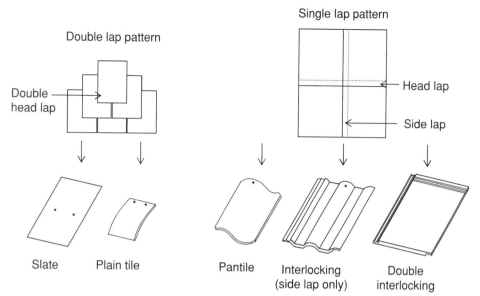

Double lap pattern

Single lap pattern

Double
head lap

Head lap

Side lap

Slate   Plain tile   Pantile   Interlocking   Double
                                              (side lap only)   interlocking

Figure 18.2 Slates and tiles – patterns and types.

joints can be used at the head. Supports also have to be close together as the slate will only be of limited size and each will require support. The thin material limits the weight of the individual slates, but because they are flat and have butt side joints they have to be laid as a double layer, as illustrated in Figure 18.2.

The other materials such as clay and concrete are permeable and therefore will have to rely on their thickness to 'store' water during rain and release it during dry periods. Tiles of concrete and clay are often shaped, not to provide any significant gain in strength but to provide weathering at the side (and in some cases head) laps so that the tiles can be used in a single layer. The single layer patterns and typical forms of tile are shown in Figure 18.2.

The joints between tiles or slates are not sealed so they can only be used on pitched roofs. The design of the side and head laps will determine the minimum pitch to which each design will work. Although relatively heavy, the tiles and slates will have to have some fixing against wind lift and must be held to prevent sliding down the sloping face of the roof.

In the UK, tile and slate roof coverings have become the dominant forms of use for small-scale structures such as housing. The generic form lends itself to the use of many locally found materials that can be economically produced for a local market. Indeed, in many areas the traditional houses are characterised by the dominance of the locally available roofing materials. With larger-scale production, particularly in concrete, the local feel is often retained in the finish, colour and possibly the shape of the tile.

More traditional materials such as stone, thatch and turf have not been discussed, but they fit into one of the three generic forms. Stone is most often used like a slate, only its higher permeability leads to thick and heavy coverings. Thatch and turf are variants on the continuous covering, but unlike the majority of currently used materials they are permeable and hence often have considerable thickness. They rely not only on 'storing' the rain till dry weather returns but also on drainage taking place within the material down the slope of the pitch with an underlay (traditionally bark) to form a surface over which any residual water can drain to the eaves of the roof.

There is a potential for reuse of old tiles and slates for refurbishment projects, but often old roofs are replaced because the tiles are weathering or deteriorating to the point where the reliability of the old roof is in question and therefore the old tiles are seen as of limited value on new work.

Interlocking concrete tiles

Other than in specific conservation areas or on more bespoke or prestigious designs, the concrete interlocking single lap tile has emerged as a common choice of weatherproof covering for new housing in the UK. It provides an economic solution, with variations in finish, shape and colour available. In many cases traditional shapes of clay and even flat stone tiles and slates have been reproduced in concrete which, while more regular than their traditional counterparts, can provide a local look to the roof covering.

The concrete tile has to be cut to form hips and valleys and a range of special tiles are needed to create the ridge and hips, including flue terminal and ventilation tiles to work with eaves ventilation to limit condensation in unheated roof spaces. Some care is required in setting out the tiles to ensure that whole tiles are used at the verge as well as ridge and eaves, as part tiles are vulnerable to wind lift, a particular risk at the verge.

The concrete single lap interlocking tile is manufactured in a variety of styles or profiles which may influence the size. They are, how-

ever, normally around 420×330 mm and weigh around 50 kg/m^2 when laid on the roof. Tiles made to look like traditional shapes, such as the pantile, will maintain the traditional size, in this case around 380×230 mm, but the overall weight will be similar. Each tile has two nibs to hang them to the batten and most have the provision of a single nail hole to fix the tile down, as shown in Figure 18.3. Where nail holes are not provided the tiles will need to be clipped against wind uplift.

The battens are normally 38×19 mm for support at up to 450 mm centres and 38×25 mm if support is only provided at 600 mm centres. For the 420×330 mm tile, battens at a gauge of 345 mm will give a head lap of 75 mm. While all edge tiles have to be nailed or clipped, the exact requirement for how many other tiles should be fixed will depend on the local exposure conditions, but some proportion of the tiles will probably need to be fixed down. Additional clipping to the front edge of these tiles, particularly at the eaves, and external verge tile clipping may be specified to overcome wind uplift forces.

Like all roof coverings, each tile type will have a minimum pitch at which the tiles will remain weatherproof, which can, to a limited extent, be varied by the size of the head lap. The style of the tile determines the minimum pitch, with some profiles being suitable down as low as 17.5° with the increase in head lap; others are only

Figure 18.3 Interlocking tiles – fixings and felt.

suitable for 30° pitch. Manufacturers' recommendations must be consulted for a particular specification including any maximum pitch. The head lap will also determine the weight of the covering and the gauge of the battens.

With the specification so far there remains a risk of water ingress. In periods of prolonged or heavy rain and wind some water may get through the tile layer. Wind-blown snow might also penetrate the laps in ways rain might not. Any damage to tiles will also lead to a loss of the weatherproofing function which could cause internal damage even in the short term during a storm, before any repair is effected. To provide a secondary line of defence against these circumstances the battens should be fixed over a layer of sarking felt. This is shown in Figure 18.3. This material is a thin impermeable sheet traditionally felt-based (hence the name) with a bitumen-impregnated hessian layer but now is more likely to be a plastic-based material. Based on polyester, reinforced polythene or spun-bonded polypropylene, these underlay materials are more durable and have greater tear resistance. They can now be obtained as a breathable material that remains waterproof but allows the movement of water vapour in order to reduce the overall vapour resistance on the cold side of the insulation, necessary in unventilated roofs. Although lapped to ensure continuity of weather protection, the laps are not sealed as the level of exposure beneath the tiles is very sheltered and large volumes of water are not envisaged. The felt sags between the rafters to form a drainage channel under the battens leading to the eaves. At the eaves, care must be taken to detail the felt at the gutter to make sure that any water that does get past the tiles can be effectively passed into the gutter. This detail is discussed later in this chapter.

It is not the intention of this text to give details of all the options for roof coverings. If other options are to be investigated it is necessary to be sure that the basic waterproofing mechanism is understood; how the covering is supported, jointed and fixed to the structure, together with some analysis of the risk of damage in the event of failure of the primary waterproofing layer.

All of this will determine the performance of the structure, the loading and the movement that might be allowable without loss of performance of the weatherproofing function of the roof. For instance, the use of plain tiles is a traditional double lap system, as shown in Figure 18.2. The tiles are 268 × 165 mm with two nibs and two nail holes for fixing to the battens at a gauge of 100 mm. Although a small tile, its double layer pattern means a loading of around 80 kg/m². The tile is cambered to help break the capillary path between the tiles; however, its size means that it still only works on a minimum pitch of 35°, but it works well vertically (all tiles nailed) as a cladding material.

Structure

As the floor was based on the joist (Chapter 17) so the pitched roof is based on the rafter. Figure 18.4 shows the basic arrangement and variables of the rafters. As with the joist for the floor, the typical size of the house generates limited spans, so the rectangular-section timber rafter is the initial potentially economic choice. However, unlike the floor the rafter has to be supported at the required pitch. For the dual pitched roof a pair of rafters provide mutual support at the ridge. With this loading and support pattern there are six possible modes of structural failure, illustrated in Figure 18.5:

- Overturning of rafters
- Thrust at the wall
- Deflection of the rafter
- Lateral torsion buckling in the individual rafters
- Failure of the rafter in bending or shear
- Wind uplift

The pairs of rafters are very vulnerable to overturning, so even on the smallest spans some longitudinal members will be required. While the tiling battens act in this longitudinal direction, specific timbers associated with the roof structure need to be provided. The simplest of these is the ridge board inserted between the two rafters running the full length of the ridge up to the gable walls. This arrangement of rafters

Figure 18.4   Timber rafters – arrangement and variables.

Figure 18.5   Modes of structural failure of rafters.

Figure 18.6 Couple roofs.

and a ridge board is known as a couple roof and is shown in Figure 18.6. The ends of the rafters at the wall are fixed (birdsmouth cut and nailed) to a wall plate. The wall plate is a timber around 100×75 mm bedded flat in mortar on the top of the wall. It provides a bearing for the end of the rafter and level support to maintain a good line on the end of the rafters for the eaves.

With an increase in span of the roof and/or slenderness of the wall the thrust soon becomes a major instability, limiting the span of the couple roof to around 3 metres. While it is possible to solve the overturning of the wall in the design of the wall itself with, say, buttressing, it is more economic to tie the feet of the rafters together with a ceiling tie. This forms a structure known as a close couple roof and is shown in Figure 18.6. The ceiling tie also acts as the ceiling joist as it provides support for the ceiling and any storage or maintenance activity in the roof. This induces bending, but the primary structural behaviour for the roof as a whole is in tension. For the structural action in tension to develop, it is necessary to fix the rafter and ceiling tie together, which can be achieved by nailing. The size of the ceiling joist is determined by the bending action and will therefore appear similar to floor joists. It is possible to reduce the size of the joist by breaking up the span with binders if suitable support can be found for the binders.

As the span or the pitch increases, the length of the rafter (and ceiling joist) increases and hence the bending forces. Deflections increase, as does the potential for lateral torsion buckling and the ultimate failure in bending or shear. Shear failure is unlikely, as the loading is relatively light and the span relatively long. This is the condition that often makes deflection the design criterion even before a bending failure is likely. To overcome either deflection or bending failure, additional depth of rafter is the most effective way to provide resistance to the bending forces. However, the width of the rafter is also significant in that it limits the tendency to lateral torsion instability. The need to increase both depth and width of the rafter increases both self-weight and cost. The economic limits for this closed couple structure with its simple pair of rafters and a ceiling tie is around 5 to 6 metres.

It is possible to identify two alternative ways to provide a structure with a longer rafter and hence provide economic roof structures over 6 metres in span:

- The purlin roof
- The trussed rafter

The purlin is a longitudinal member that provides intermediate support to each rafter as shown in Figure 18.7. The purlin in turn has to be supported, the frequency of which will determine the span of the purlin and hence its cross-sectional size. The purlin is more heavily loaded than the rafter and can therefore only provide support over relatively short spans if the purlin is to be constructed with a rectangular timber section. While rafters are likely to be around 50 × 125 mm, the purlin is likely to be around 75 × 200 mm and will require support, at between 1800 and 2400 mm centres. This will require the use of an internal load-bearing wall to provide a support for the struts from the purlin, as shown in Figure 18.7, although the ends of the purlin can be built into a gable wall.

It is possible to use either a steel beam section or a deep timber purlin made from a composite timber member in a 'box', 'I' or truss form to span between gable walls and hence eliminate the need for the internal load-bearing wall. These options are also shown in Figure 18.7. While the purlin in steel would remain a relatively small section, the composite timber solution can become quite deep, making this an option that restricts movement in the roof space. Another option to eliminate the internal load-bearing wall is to use trusses at every fourth rafter to support the purlin with common rafters supported by the purlin between. This adds expense and again limits the space available in the roof. The use of the truss and common rafters is a very traditional solution used for open barns and halls since medieval times but does not prove economic for houses.

However, in more recent times the idea of the truss has been revived in the introduction of the trussed rafter. This has been made possible by the development of the gang nail plate to make the joint between members. This is illustrated in Figure 18.8. It provides a fixing that allows butt jointing of the timber members and has the ability to develop sufficient tensile strength in the joint to construct a lightweight pin-jointed frame (see Chapter 11) for each rafter. Although it is possible to fix gang nail plates on site, they gain their full economic advantage in factory assembly processes even with the additional transport costs of delivering the trussed rafter as a whole.

The structural efficiency of this system leads to slender sections. This creates the need to not

Figure 18.7 Purlin support alternatives.

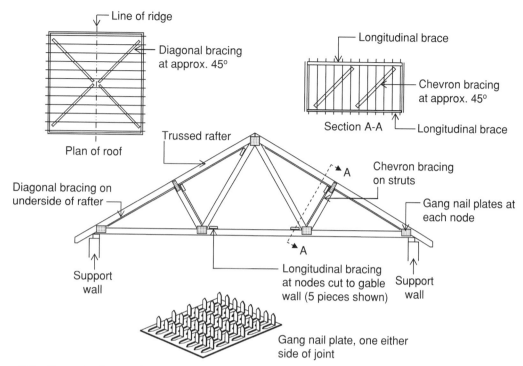

Figure 18.8   Trussed rafter roofs.

only handle the trusses with care during transport and assembly but also reassess the stability of the roof structure as a whole. It is not possible to use a conventional ridge board and the slender timber sections are more susceptible to lateral distortion, particularly those in compression. This leads to a requirement for a number of longitudinal and diagonal bracing members as shown in Figure 18.8.

Even with the need for care in handling and the additional bracing, the trussed rafter currently provides the most economic solution. Its popularity has led to a wide range of components to allow hip and valley roofs to be constructed and alternative truss configurations to allow rooms to be built into the roof space. Standard designs for roof spans of 12 metres and over are now available and most manufacturers offer to design and manufacture specialist design options.

Having specified a structure to support the finish to ensure its performance from the vertical loading it is necessary to consider the proposed solution under wind-loading uplift conditions. If the uplift is sufficient to overcome the dead load then failure is possible. Given that the fixing of the finishes is sufficient to resist their separation from the structure, the effect will be initially to relieve the load and then, with increasing wind speed, the lifting of the roof as a whole. If this is a possible design condition then the roof structure must be strapped down to the walls. This is a common condition for the flat roof where lightweight coverings are used and strapping is an essential detail. With a pitched roof, the weight of the finish makes uplift less likely in all but the most onerous UK wind conditions. Where strapping is required the straps should be at 1500 mm centres and be at least 1000 mm long, with the bottom fixing to the wall in the last 150 mm of the strap. The feet of the rafters need to be located where nailing or galvanised steel brackets can be used to fix rafter to wall plate. If nailing is used then care has to be taken not to damage the gang nail plate and reduce its structural effectiveness.

Figure 18.9 Strapping of external walls to roof.

If the house has a gable end configuration the gable wall is also subject to these wind forces along the edge of the roof, which exerts sideways forces likely to damage the top of the wall. The top of the wall is adjacent to a fully braced and stable roof structure so it is possible to strap the top of the wall to the roof structure to stabilise it against these horizontal forces. These two sets of strapping, roof to wall at the eaves and wall to roof at the verge, are shown in Figure 18.9.

It is worth noting that if uplift is suspected, and the roof structure is successfully fixed to the walls, then uplift will reverse the action on the structural members and joints. Joists on flat roofs and rafters on pitched roofs will reverse bending. The purlin roof, whose joints rely on compression, will be lifted apart. The truss will have upward forces, reversing the forces in many members. If this is suspected then members in tension under normal loading may become vulnerable to lateral instabilities, as they will not be braced for the compression forces that develop in high winds.

Internal finishes and other elements

The internal finish associated with the roof is a ceiling finish the same as that below a floor and

the solution is likely to be the same. Analysis of the ceiling joists or the bottom chord of the truss shows a background the same as the floor, as discussed in Chapter 17 under the same heading of 'Finishes and other elements'. The solution of a plasterboard finish to receive paint is therefore likely to be the same choice.

The early part of this chapter introduced the idea of solar energy collection devices becoming part of the design. These will have to be connected to the roof structure and their loads need to be taken into account in the design. These fixings and the services connections will penetrate the tiling, requiring careful detailing if the weatherproofing is not to be compromised.

Thermal performance

The proposal so far provides waterproof and stable construction with a finish to the ceiling for rooms below. It gives very little resistance to the passage of heat and would, as so far considered, prove to be a large proportion of the overall heat loss from the building. This can be overcome most effectively with the addition of insulation.

This is best provided at ceiling level so as not to heat the roof space, so long as it is not being designed to be habitable. This will require any water services or tanks in the roof to be 'lagged' in their own insulation materials to prevent freezing in cold weather. With the introduction of insulation at the ceiling level, the roof space may not be receiving heat from the living areas below, but it will still be receiving water vapour. This cold roof increases the risk of condensation on the timbers and on the underside of the sarking felt. On the timbers this may be a durability risk, while on the sarking felt the condensation may drip back onto the insulation. While it may seem sensible to specify a vapour control layer below the insulation, the detailing and installation difficulties make the provision of an effective layer unlikely. The alternative approach to condensation control, ventilation, is more likely to be reliable.

This ventilation must ensure a good flow of air throughout the roof. The wind will provide a cross flow if ventilators are introduced at eaves level. However, to ensure a full flow of air right up to the ridge level it may be necessary to introduce ventilation at the ridge or in individual tiles near the ridge.

Insulation is likely to be placed between the joists on the ceiling plasterboard and can therefore be in the form of bats or quilt or even loose materials. If insulation in excess of the depth of the joist is specified then access into the roof space can be difficult as the safe areas to tread on the joists might be obscured.

Life cycle considerations

The construction of the roof considered so far includes concrete, timber and metal components. These components are subject to various exposure conditions, but none gains the full protection of the internal environment. Clearly the tile finish is the most exposed, where the concrete components are probably most vulnerable to frost attack and colour degradation from sunlight. The tile is fixed with nails and these will be subject to corrosion and are therefore normally specified in a non-ferrous metal. Aluminium alloy or copper nails are suitable for normal use, but in aggressive environments silicon bronze or stainless steel may be specified.

The roof finish will experience a wide range of temperatures on an annual basis. Exposed directly to the sky, it will suffer from surface temperatures more extreme than the air temperature due to radiant exchange, reaching high temperatures under the mid-day sun in the summer and low temperatures under clear night skies in the winter. This will induce dimensional changes that have to be accommodated without loss of performance. While this is of concern with continuous finishes, the number and type of joints and fixings on tiling and slating allows individual tiles to move over each other, limiting the chance of stress developing and leading to failure of the components.

Perhaps the most exposed timber component is the tile batten and this should be specified to be treated softwood. Again its fixing nails should be a non-ferrous metal. The sarking felt will be protected by the tiles from most sunlight but will tend to become more brittle with age. This makes it most vulnerable where it is bent, particularly at the eaves where it is dressed into the gutter. While this is not so much of a problem with most of the plastic-based underlay that will be UV stabilised, eaves protection strips should be used with more traditional felts.

The main truss is now protected from the direct effects of the weather and, with adequate ventilation for the roof space, free from the risks of condensation. The roof space is, however, open to insect attack, the most severe in the UK being the house long horn beetle. This insect is limited to only a small area in the UK but has the potential for such damage that roof timbers in these areas have to be preservative treated. The trussed rafter gang nailed plate is made of galvanised steel, but the 'nails' are punched out after galvanising. While the softwood should not interact with the exposed steel nail edges and ventilation will limit condensation that might promote corrosion, buildings in humid or exposed conditions, particularly coastal areas, might give concerns for the long-term corrosion of the plates. If this is the case then it is possible to specify stainless-steel plates or some other additional protection.

The ceiling construction will experience internal conditions and will not be exposed to a high risk of damage. Some forms of insulation above the ceiling may suffer from bedding down, reducing the insulating effect, but should not be subject to major deterioration mechanisms given the conditions within the ventilated roof space.

Wind-blown dirt will be stopped to a great extent by the felt under the tiles, but roof spaces do become dusty places. This will not inhibit infestations of birds and insects and even rodents that can climb using drainpipes and plants growing on or close to the walls. Some care has to be taken in detailing, particularly at eaves, to limit access if infestations are to be avoided.

Modern concrete tile roofs have a life expectancy of around fifty years. In this time the roof finishes and structure should need little regular maintenance, perhaps the repointing of ridge or hip tiles. The major maintenance is likely to be around the eaves depending on the choice of materials. Following extreme weather events, particularly high winds, minor damage may occur so access will have to be arranged to undertake the repair. It is unlikely that there will be any provision for access to the roof in the original design, relying on the need for temporary works access and health and safety measures to be provided at the time of the repair. However, it is possible that during the repair and maintenance activity, individual tiles will be stepped on and therefore they must have sufficient strength to span between the batten and the tile below on which it will be resting. This does need to be foreseen in the design of the tile and will have to be one of the standard tests on tiles.

The fifty-year life means that replacement of the finish if not the whole structure is likely during the life of the building. Once a safe environment is established for working at height and the old roof is removed the replacement is the same procedure as the original installation, other than the house being in occupation during renewal. Tiles may be sorted, with some being saved for reuse but most will be weathered with surface deterioration so could be crushed to be recycled as clean fill. The condition for the battens is likely to be very variable, but a range of recycling options might be available rather than sending to landfill for disposal.

Edge and junction detailing

The range of configurations introduced at the beginning of this chapter indicates the need for a range of edge and junction details to be available, not all for one building, but the components and details need to be sufficiently developed to provide a reasonable choice for a wide range of buildings. Details and junctions are required at edges such as eaves and verges as well as where roof surfaces meet at ridge, hip, valleys and internal gutters and at abutments where a roof meets a wall. These are then needed in some combination for penetrating structures such as a chimney or dormer window.

It is not the intention of this text to detail all of these but to take a range that will illustrate the key features of these assemblies. The following will be discussed:

- Eaves (edge with rainwater collection)
- Verge (edge without rainwater collection)
- Ridge (meeting of surfaces at top of roof)
- Abutment (junction of roof with wall)

Figure 18.10 shows a typical eaves detail for a roof with concrete interlocking single lap tiles on a structure based on the trusses rafter with insulation above a plasterboard ceiling. Perhaps the first decision concerns the size of the overhang. This most obviously affects the appearance but also the weathering to the top of the wall. Traditionally buildings in areas of high exposure will have wide overhangs, but even in less-exposed conditions eaves with little or no overhang are vulnerable to damp particularly as the age of the roof increases.

The detail is shown with an overhang of around 300 mm with both a facia and a soffit board. The facia board, while providing a visual edge to the roof, gives a surface to attach the gutter to and a continuous straight and level support for the edge of the tiles. The top of the facia is fixed above the line of the rafters to lift the edge of the tile. All the tiles on the roof,

Figure 18.10 Eaves detail.

except the one at the eaves, rest on the tile below and the facia has to mimic this line of support to keep the pitch of the tiles the same all the way to the eaves. This creates a step behind the facia that would allow the sarking felt to sag, creating a potential for ponding behind the facia rather than a clean discharge of water into the gutter. The detail in Figure 18.10 shows a timber fillet to provide a solid surface for the felt that is taken over the top of the facia board and dressed into the gutter. The tile then overhangs the facia to discharge the rainwater into the gutter. Tiles with a bold or deep profile then present relatively large gaps above the facia board, giving potential access for birds that may damage the felt. Some filler piece, fixed to the top of the facia, may be necessary to eliminate this risk.

The facia has a primary role in weathering the eaves and providing effective discharge of the rainwater into the gutter. The soffit board hides the ends of the rafters and seals the eaves. While providing protection from birds and the larger insects, it also has the effect of limiting airflow into the roof space, which is required to provide

the ventilation to minimise the condensation risk on the cold side of the insulation. The detail in Figure 18.10 shows a ventilator associated with the soffit board and a tray fixed to the top of the rafters over the insulation to ensure a path for the air into the roof space. Taking the insulation right into the eaves is important in order to limit cold bridging at the junction of the wall and the roof.

Both the facia and soffit board are shown constructed in timber that would have to have an applied finish, usually paint. Although the whole detail would be considered as exposed, the facia, and particularly the soffit, are relatively well protected, but maintaining these boards still has to be considered. This has led to a number of alternatives being used for the boards, often plastic, but the initial cost and ease of fixing makes the timber solution the common choice for new houses.

The transfer of load from the roof structure to the wall also takes place at the eaves and therefore has to be resolved in this detail. Each trussed rafter rests on the wall plate. This is a timber,

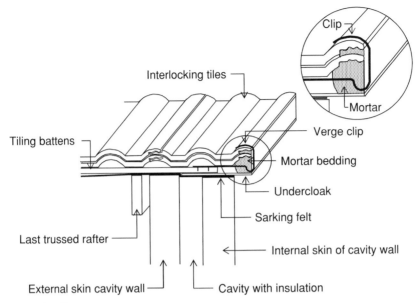

Figure 18.11 Verge detail.

bedded level on mortar, on the brick closure that seals the top of the cavity. This provides not only a level surface to ensure the line of the edge of the roof but also an even transfer of load to the wall. In Figure 18.10 the wall plate and closure are on the inner skin of the cavity wall, effectively transferring the roof load only to this skin of the wall. This will have to be taken into account in the choice of the detailing and materials of the wall considered in Chapter 19. For the sake of clarity, Figure 18.10 does not show the fixing of the trussed rafter to the wall plate or any strapping, which would be required to hold the roof against wind uplift provided at the eaves, as illustrated in Figure 18.9.

Figure 18.11 shows a typical detail of a verge over a gable end wall. The detail shows a bedded verge with a 50 mm overhang without barge or soffit board. The bedding is cement mortar, a wet process to seal the edge of the tiling from water ingress. Some tile systems are manufactured with special overhanging tiles or other covering components to provide a dry verge. These have been developed to overcome the long-term maintenance potential of the cracking and loss of the bedding material from between the tiles.

It is possible to provide an overhang detail similar to the eaves, but this requires some additional structure to provide support for the tiles over the wall. The problem arises because the last effective fixing to support the batten is the rafter on the inside of the wall positioned some 50 mm from the wall. If the tiles are to be finished with only a small 50 mm overhang on the external wall, then the battens extended over the wall can provide sufficient support. If a larger overhang is required at the verge then a gable ladder has to be fixed to the last rafter and extended over the wall to provide support to both a bargeboard (the equivalent to the facia board) and a soffit board. A bedded or dry verge detail can then be chosen but this time with the 50 mm overhang formed over the bargeboard not the brick wall.

The main detail in Figure 18.11 shows the last rafter with the sarking felt taken over the wall and the batten extending about halfway across the external skin to provide support for the last tile. In this detail the brickwork of the external skin has to be cut to a good line to provide a surface on which to lie the undercloaking. This is a thin rigid board, often a fibre reinforced cement material, positioned between the felt

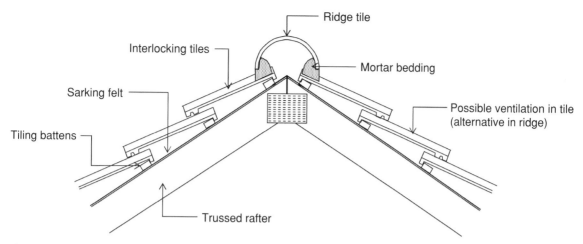

Figure 18.12   Ridge detail.

and the batten. This undercloaking provides the clean visual line to the top of the wall but also the support for the mortar material, so it must be a dimensionally stable material to maintain the integrity of the seal to the edge of the tiles if the bond with the mortar is lost.

To give extra protection to the tile against wind uplift, clips are fixed to the battens and then the tiles are bedded and nailed, with the bedding material struck and finished to give a clean edge to the tiling. To help with the setting out and to avoid cutting, half tiles are available, and the overhang can be varied between 38 and 50 mm. With some profiles, left-hand tiles are produced without the left-hand (upward-facing) interlocking grooves that would otherwise show at the verge edge.

Figure 18.12 shows a typical ridge detail. Like the verge it shows a traditional bedded ridge, although dry ridge systems are available for some tile profiles. The figure shows a half round ridge tile (without nail hole) that is bedded on mortar on top of the tiles that will be nailed to the batten. Each ridge tile is 450 mm long with butt joints that have to be solidly bedded at the joint to maintain the waterproofing and provide a full fixing. If the profile of the tile has a deep role then the thickness of bedding may become excessive and liable to slumping and shrinkage. This is prevented with small pieces of tile known as dentil slips inserted into the mortar and norm-

ally showing as a feature on the tiles at the ridge line. If ridge-level ventilation is required then special ridge tiles can be used. The sarking felt, normally continuous over the ridge, will have to be cut and dressed up around the short internal duct that connects the ventilation terminal with the inside of the roof space.

The last detail, the abutment, is shown in Figure 18.13. The additional components in this detail are the flashings. Normally formed in lead, they seal the junction between tiles and wall. For tiles with a deep profile typical of the concrete interlocking single lap tiles illustrated, it is usual to use a cover or overflashing where the lead is dressed on top to the tile, over the roll on the tile and onto the natural drainage path formed by the tiles profile. Where tiles do not have a deep profile, as with plain tiling, then soakers or a secret gutter detail will have to be used (not illustrated).

However the flashing is formed in the plane of the tiles, the flashing has to be sealed against the brickwork. This is achieved by stepping the top of the flashing and turning the horizontal edges into the brick joints that have been racked out to a depth of 20 mm to receive the lead which is then held in with lead wedges and then pointed to match the rest of the brickwork. If the wall facing material does not provide natural joints to turn the lead into, then a groove has to be provided.

Figure 18.13 Abutment detail.

Under the lead flashing the felt is turned up the wall and the tiles are butted up to the wall. Tiles may be cut to maintain setting out if necessary as long as the tile is fixed and the flashing covers the roll.

Production operations

Roofs have to be constructed at height and therefore there is a need for a safe working platform for the operatives to prepare and fix the materials and components into the final assemblies. There then needs to be safe access for both operatives and the materials and components.

For the roof work to start, the walls will have to be complete up to the wall plate. Strictly, only the inner skin needs to be complete. In a masonry wall it is not sensible to build the skins independently, although this is not the case in timber-framed construction where the whole structure is complete before the outer skin is built. The scaffold used to construct the wall should be in place, giving safe access and a working platform for operations completed at the edge of roof.

The basic sequence of work is to erect the structure, complete the timber construction at the

eaves and verge and then undertake the tiling. The ceiling will be completed along with the other finishes. Insulation will have to be installed after the ceiling is fixed.

All the components will be delivered to site with some level of preparation. Trussed rafters will be factory assembled, but the ends of the rafters will have to be cut to suit the eaves overhang detail. Other timbers such as fascia, barge and soffit boards will be delivered in lengths to be cut to fit, as will the tiling battens that will be factory pressure impregnated with preservative. The felt will be delivered in rolls as will any lead for flashing. Tile and guttering systems will be delivered with any special components and fixings, but the main components (the tiles and lengths of guttering) may still need to be cut on site to maintain fit and layout for some details.

All these components will need to be handled and stored on site. The trussed rafters are perhaps the most vulnerable. Excessive stress can be put on the joints during storage but particularly during handling. If the truss is manhandled, lifting should be at the eaves joint with the truss in the vertical plane. This ensures the stress in the truss, as it will be the same as when it is supported in the final position. This is also the case for storage in the vertical position where supports should be level and at the eaves joints, although flat storage is possible if care is taken in lifting back to the vertical plane. Mechanical lifting often cannot be from the eaves joints but should always be arranged to be at the structural node points with a guide rope if a crane is used.

With the other components, damage and waste should be avoided with good storage conditions and careful handling. In particular, the hazards of falling from the roof and designated areas for machines transporting and lifting the components should be considered in planning site procedures.

Flat roofs

Given the case study approach being used in these chapters it is unlikely that flat roofs will be used on a house in the UK at the beginning of the twenty-first century, so they are not going to be dealt with in detail in this text. While the flat roof possibly offers a lower initial cost, this is only achieved with the less expensive continuous roof coverings such as three-layer felting. Other continuous coverings are more expensive and make the overall flat roof more costly than a pitched equivalent. The three-layer felts have a lower life span of around 15 to 20 years and are more susceptible to damage due to movement and general exposure, making them less reliable and a high maintenance cost for the owner. Any lower initial cost is, therefore, a false economy if funds are available at the time of construction.

However, a well-detailed flat roof with an appropriate specification can provide a workable solution. As indicated at the beginning of this chapter, the flat roof has a joisted structure similar to a floor but has to be provided with a fall to ensure effective removal of rainwater into the gutters. This is achieved by either sloping the joists (giving a sloping ceiling) or placing tapered battens known as firring pieces on top of the joists. The boarding, to receive the waterproof finish, should be specified to be damp resistant. Some protection can then be given to the surface of the waterproof covering against direct weathering and surface temperature effects, with a white reflective topping layer often provided by light-coloured chippings.

Insulating flat roofs has to take the condensation risk into account, as the waterproof layer is also a vapour-resistant layer and will effectively seal all vapour in the construction. If the insulation is laid between the joists, a 50-mm ventilated gap above the insulation is essential. The difficulties of fixing and edge detailing a vapour control layer under the joists will make it relatively ineffective. The roof with insulation between the joists is known as a cold deck roof as the boarding is on the cold side of the insulation. An alternative, known as warm deck solutions, can be adopted where the insulation is above the boarding. Now a vapour control layer laid on the boarding can be made more effective and the joist space can be fully ventilated. A solution with even less risk of condensation is known as an inverted roof where the insulation is above the weatherproofing layer. The entire construction now remains warm and the condensation risk is eliminated. However, the insulation is now outside, so its water absorption and impact strength for maintenance loading will have to be considered, limiting the type of insulation material that can be used.

Summary

(1) The pitch and the span of the roof dominate the choice of both roof coverings and structure.

(2) The roof will need insulating to resist the passage of heat and may have some fire resistance requirements for the coverings. If used as part of habitable space then lighting with windows and roof lights may be required.

(3) The pitched roof is the most likely option for this case study as this represents the traditional look of the UK house and recent problems with flat roofs have led to the pitched roof being seen as the more robust construction.

(4) While continuous and large sheet coverings can be used, it is tiles (and slates) that have dominated the traditional pitched roof in the UK, with the concrete interlocking single lap tile being the most economic solution in the case study.

(5) With the variety of colours and profiles, some based on traditional tile shapes, together with their suitability for a variety of pitches, the concrete interlocking tile can be used in most circumstances.

(6) The structure is based on the timber rafter. There are a variety of ways to hold the rafter safely at the required pitch for a range of spans typical of the house in the case study. While solutions based on the purlin have been used in the past, the trussed rafter provides an economic solution as long as the required bracing and handling and fixing details are followed in the production process.

(7) The fully braced roof structure provides a background for a plasterboard ceiling finish and support for the external walls at gable ends. This structure may still be vulnerable to wind uplift and may need strapping down to the walls at the eaves.

(8) Insulation will most probably be provided between the ceiling joists where the cold roof space can be ventilated to reduce the risk of condensation.

(9) The roof finish is unlikely to last for the full life of the house and will need to be replaced. Detailing will have to ensure conditions that maintain durability and barriers to infestation and excessive dust and dirt that may enter during the life of the building.

(10) The variety of roof configurations to suit all building plan shapes requires a number of edge and junction details. Details have to be devised for eaves, where water is discharged into gutters, and verges, the edges without rainwater collection. Details also have to be chosen for ridges, hips and valleys where two roof surfaces meet and at abutments, the junction between a roof surface and a wall.

19 Walls

This chapter considers external cavity walls, separating (party) walls and internal partitions. It introduces their performance requirements and provides the analysis of their construction in both masonry and timber frames, leading to typical details and specifications for each.

Wall types by function

Walls will play a major part in the construction of the house in the case study, providing a wide range of functions dominated by the structural and environmental roles they are asked to perform. Walls are given different names depending on the role they play in the overall design based on the broad set of functions they perform.

We talk about loadbearing and non-loadbearing walls normally based on the direct analysis of the load paths from dead and imposed loads. This is a limited definition of being part of the structure, as overall stability associated with wind loading and component instability often relies on walls that may not be carrying the direct loads from the construction and the occupation of the building. Many walls that are called non-loadbearing are part of the structural system and therefore are subject to stress under certain conditions.

Environmentally we talk about external walls, separating (party) walls and partitions (internal walls), each having an associated set of functions that will determine the choice of construction. Given the exposure conditions in this case study, external walls are dominated by the needs of weathering and insulation to resist the passage of heat. Openings for windows and doors are a major influence on the design in so far as they fulfil functions of access, light, views, security and ventilation required of the wall. However, the creation of the opening is complicated by the need to maintain the function of the wall as a whole. The components associated with windows and doors will have limited performance in strength and insulation, functions that were all carefully incorporated into the basic wall construction.

Separating walls are between dwellings in terraced or semi-detached designs. They will have no openings and no weathering requirements but will need to have functions associated with security and privacy in the need for fire and sound resistance. The other internal walls within the dwelling are known as partitions. There is a need for openings, and the dividing function will again be dominated by privacy and security. Acting as a visual barrier will not be sufficient as the wall will need some fire and sound resistance and must be robust and capable of supporting finishes, cupboards and shelves and integrating services.

General forms

We also refer to the general forms of walls in a number of ways. In housing, one distinction commonly made associated with weatherproofing is reference to the external cavity or solid wall (not to be confused by the two skins sometimes used in a separating wall for sound resistance where the 'cavity' will be designed using different criteria). The external cavity wall has gained almost universal acceptance, although the need for increased insulation and other energy-saving ideas and perhaps a wider acceptance of rendering as an external finish have revived interest in the solid wall, but it has yet to become widespread in practice.

For either cavity or solid walls, three generic forms can be employed in the construction of walls:

Cavity

Blockwork

Brickwork in
stretcher bond

Figure 19.1   Masonry walls in
brick and block.

- Masonry
- Panel (framed or solid)
- Monolithic

The basic distinction is in the size of components and their spatial arrangements which determine the types of materials, jointing and production process involved. All walls need to provide a continuous form. If they are loadbearing walls they need to be capable of distributing loads evenly throughout the structure. They also need to provide continuity for environmental control plus the capacity to have openings formed within the construction.

Masonry is characterised by the heavy materials such as stone, fired clay or concrete that lead to components such as bricks and blocks that are normally relatively small compared to the dimensions of the wall and often regular in shape to form the 'bonding' patterns that give stability and continuity. The choice of materials and the size and regularity of the components vary greatly in their properties, appearance and cost. Given their weight, locally available materials have in the past been chosen on economic grounds based on the cost of transport and this has led to geographically recognisable styles. With modern manufacturing methods and the relatively low cost of transport, fired clay bricks and concrete blocks are the dominant choice in most areas of the UK. An external cavity wall constructed with a concrete block inner skin and a brick

external skin is shown in Figure 19.1. The bricks and blocks have a modular size that includes the 10-mm mortar joint and that allows the bonding patterns (stretcher bond shown) and the use of bricks and blocks in the same wall. This basic relationship is shown in Figure 19.2. For the brick this modularity is extended to its third dimension. The modular width is 112.5 mm, half the modular length, which allows thicker walls to be built by laying the brick with the header to the face of the wall. This full modularity of the brick is shown in Figure 19.2. Blocks are not modular in the third dimension, being available from 75 to 200 mm thick.

Local availability may still influence choice in conservation areas and now local materials with low transport needs can reduce energy use (rather than cost) and can therefore contribute to sustainable development.

Panel walls are storey height construction that can be constructed in position or prefabricated in panels in modular widths to be combined into walls of various lengths to suit the design and the site constraints. Framed panels in either timber or lightweight steel sections form the basis of many prefabricated systems that can be ordered from the factory and erected and jointed on site. The basic form of a framed panel wall in timber is shown in Figure 19.3. Solid panels can also be prefabricated, normally associated with concrete but more recently developed in sandwich construction where high insulation values are

Modular height 75 mm
Working size 65 mm

Stretcher face

Header end

Modular length 225 mm
Working size 215 mm

Modular width 112.5 mm
Working size 102.5 mm

Block modular
length 450 mm

Concrete block size
440 x 215 mm

Block modular
height 225 mm

Brick size
215 x 65 mm

Brick modular height
(gauge) 75 mm

All mortar joints 10 mm

Brick modular
length 225 mm

Figure 19.2 Masonry brick and block modularity.

Top rail

Board on one
but normally
two sides

Studs at 400 or
600 mm centres

Bottom rail

Figure 19.3 Panel wall in timber frame.

achieved without cold bridges, by forming rigid insulation materials between timber composite board materials. Particular attention has to be paid to jointing solid panels.

While panels form the basis for some of the most advanced factory-produced prefabricated systems for housing they have in the past been fabricated on site and lifted into position, and this may still have some merits in low set-up and transport costs but is not now often adopted.

The 'component' of the monolithic wall, as the name suggests, is the whole wall of the house. This has to be an in-situ process where any joints that are necessary are dictated by the pace of the production process. These are known as day joints as they are determined by how much wall can be built in a day, but when cured the wall must act as a continuous material and this may also dictate the position of these joints. Now made of concrete, with the use of temporary formwork and false work, monolithic walls have in the past been associated with soil construction. Some soil walls used formwork (pisé), although this was not always necessary where higher clay contents were found in the local material (cob) where the wetter clay/straw mix held the material together during construction. Currently, monolithic walls are rarely used in the UK, but again some of the environmental ideas, particularly using walls for thermal capacity and with the low embodied energy associated with soil construction, may suit this form of construction.

Walls in the case study

For the house in the case study the initial suggestions to be considered will be:

- External wall – cavity with masonry external and internal skins
- Party (separating) wall – in masonry
- Partitions – in masonry or lightweight studwork

After the analysis of this masonry solution the use of timber frame will be considered, as this is a major alternative in the UK at the beginning of the twenty-first century.

Identifying which of these walls are loadbearing or more generally acting as part of the structural system can only be achieved for a particular design. This case study is not that specific, but it is reasonable to assume that the external walls will be loadbearing and at least some of the internal walls will have to carry loads from upper floors and this may well include the separating wall. These and possibly other walls will then have to take the racking forces from wind loading.

Masonry cavity external walls

The cavity wall was developed to limit the risk of damp penetration, which could be a problem with solid walls in exposed conditions particularly via the joints, aggravated by poorly maintained pointing and with the introduction of strong cement mortars. The principle was to provide an outer skin that limited but did not have to eliminate moisture penetration under all conditions. The cavity, a clear air space, provided the guarantee of damp-proofing as long as the cavity was not breached. The inner skin then remained dry and could be used to support the finish without fear of damp within the building.

The simple principle of the cavity has proved very robust so long as attention is paid to certain details in the wall and at openings (see discussion later). A number of materials can be used for the outer and inner skin, but for the case study the initial suggestion is to choose brick for the outer skin and concrete block for the inner skin. From experience the thickness of the internal and external skin can be expected to be around 100 mm. Given the scale of house construction this would seem to be a reasonable estimate, but the actual size will have to be determined by the performance requirements and the materials chosen. If the thickness deviates far from this size then the additional material will have to be justified from additional performance or economy and details will have to be reassessed. The width of the cavity is the other key part of the specification of this form of wall. The cavity only works if the width of the cavity at all points can be maintained within tolerances and kept clean. These are both production problems so the production process dictates a minimum size for the cavity. These are, therefore, related to the materials specified for each skin; in this case both are masonry for which a 50 mm clear cavity would be the usual proposal. It is not possible to provide an analysis to check this width until the whole wall specification is chosen. The size will also be determined by the introduction of insulation so this also has to be established before the size of the cavity can be determined. It will be necessary to return to confirm this initial suggestion of 50 mm when the wall specification is more advanced.

The external skin

The initial suggestion for the external skin will probably be based on its appearance. Clearly this not only will be determined by the style and status of the building but also will be a major factor in the building's response to its context. It may even be determined directly by the client who may have very strong views on what the house should be made of. Dominating the UK market is brick, although locally stone may have been historically used so reconstructed stone blocks have been adapted for cavities. Renders may have been commonly used where local quality bricks were not available or in extreme exposures such as coastal areas, and this again can be easily adapted for the cavity wall where concrete blocks can be used in the external skin as the backing for the render.

Exposure is a key factor in selecting the material for the external skin, less for damp penetration as the cavity has this function but for durability and potential maintenance. If masonry is to be used then the exposure to rain at times of low temperatures will make frost action a significant risk. Frost resistance will have to be considered not only in the brick but also in the choice of mortar. Other risks of change over time include discolouration. This change in colour may be desirable as weathered materials are often more mellow and attractive than when new. However, some discolouration of, say, pastel-coloured renders may not stand the test of time in the UK climate. Detailing to ensure limited pattern staining with good even washing from the rain needs to be considered.

Overall, the facing brick provides a good range of options, most of which can achieve frost resistance and have a wide range of colours. If local materials are to be preferred then the choice may be more limited. The modular size of the standard UK brick and block was illustrated in Figure 19.2. The brick itself is 102.5 mm thick so if laid, stretcher bond can form an outer skin, around the 100 mm originally estimated. This 'half brick' skin will limit moisture penetration to the cavity but will not eliminate it, so detailing in the cavity will have to take into account the presence of some water on the inner face of the brickwork outer skin.

The structure and the internal skin

It is not possible to proceed to accept this specification for the external skin that was based primarily on appearance, or start to think about the internal skin without being clear as to how the loads are distributed across the wall. This will be determined by the detailing from the connections of the floors and roof. Figure 19.4 shows the cavity wall with these details (discussed in the two previous chapters) to identify the dead and imposed loads together with the wind loads. Inspection shows that the internal skin takes the dead and imposed loads. No strength requirements for these loads have to be considered for the external skin so the facing brick should be sufficient. Durable bricks will have reasonable strength so the external skin should be able to support its self-weight and deal with any local crushing from the lintels over openings. If large openings are included in the design then the strength of the brick may become an issue.

It is clear that one of the key aspects of the specification of the block for the internal skin will have to be strength. The most highly stressed areas of the internal skin will be between the ground floor windowsills. Given a reasonable distribution of windows in both size and spacing apart and taking the skin to be 100 mm thick, it should be possible to use aerated concrete insulating blocks at 2.8 N/mm^2 (these may have a minimum thickness of 115 mm) for dwellings up to two stories high. These blocks are normally more expensive than blocks made with lightweight aggregates that are denser and therefore offer less insulation value but more thermal capacity. If high thermal capacity is required then

Figure 19.4 Loads on the external cavity wall.

dense concrete blocks may be a better choice as they are less expensive, certainly strong enough, but will have poor thermal insulation properties. It is therefore best to leave the final specification of the block until the thermal behaviour of the wall is considered.

The structure – overall stability

Returning to Figure 19.4, another observation would be that the cavity wall is slender, making it fundamentally unstable. Given that there are adequate foundations, two other instabilities can be identified. The first is overturning by wind forces and the second is buckling of the internal skin under compressive forces from the dead and imposed loads.

The mortar jointing and the pattern of bonding in both the bricks and the blocks ensure that the wall acts as a panel. However, as presently discussed, the external and internal skins are independent. Wind loads on the external skin are not transferred to the internal skin and the external skin does not help the buckling of the internal skin. Mutual structural support is required. This is normally provided by wall ties (although under most conditions they will probably act as struts). The form of these ties and the spacing required are shown in Figure 19.5. The width of the cavity will determine the type of tie that can be used, with the vertical twist being the only form used in wider cavities over 100 mm. On all types of tie there is a drip, a device to stop water travelling horizontally, bridging the cavity across the wall tie causing damp in the inner skin. Ties would now normally be specified in austenitic stainless steel although galvanised steel and plastic forms have been used. With the introduction of wall ties the wall now acts as one panel for both overturning and buckling but is still unstable.

Any return walls will act as buttresses to resist overturning. If the walls of the house form a cellular pattern on plan then they will receive mutual support. If the walls are predominantly in one direction as in a form known as cross wall construction then piers will have to be formed, normally at both ends of the cross wall.

Butterfly type tie

Double triangle type tie

450 mm

900 mm

Vertical twist type tie

Note drip feature at the centre of all ties

Figure 19.5 Cavity wall ties – types and spacing.

This leaves one potential area of wall vulnerable to overturning. The wall forming any gable ends will not benefit from the return walls to give stability against the wind forces. A chimney may form a pier, but in the absence of any masonry support the wall should be strapped to the roof to gain additional stability. These details are given in Chapter 18. Straps should be provided at the top of the gable wall and then at not more than 2000 mm centres down the slope of the roof.

It is also necessary to use the support of the upper floors to reduce the effective dimensions of the wall against buckling. This is achieved by providing joist hangers with restraint or strapping, details of which are given in Chapter 17. Straps should be placed at not more than 2000 mm centres for walls over 3000 mm long.

Rising damp

The cavity provides resistance to penetrating damp from rain falling directly on the surface of the wall. The materials chosen are permeable so moisture can enter the building by rising from the ground where capillary action within the materials will draw water into the bricks and blocks, creating a path for damp. The solution is to provide a damp-proof course (DPC) across each skin. In a cavity wall these DPCs are independent and can therefore be at different levels.

For the external skin the DPC has to be at least 150 mm above ground level. This protects the wall from rain falling on the ground which will splash back against the wall. For the internal skin the level depends on the internal floor level and the type of floor construction. For a suspended ground floor the DPC must be lower than any part of the floor construction that would be vulnerable to damp, while for a ground-supported slab the DPC must be continuous with the damp-proof membrane (DPM). This is shown in illustrations taken from Chapter 17 given in Figure 19.6.

For these DPCs a variety of materials can be chosen. In the past materials such as slate and even bricks with very low permeability known as engineering bricks have been incorporated in the wall, but thin, flexible sheet materials that can be laid in the bedding joint are now used. Metals such as lead and copper can be used but are expensive. The less expensive bitumen DPC, and in particular the polymeric (bitumen polymer and pitch polymer) sheets, are the most widely used. The polymeric sheets are also used for other DPC applications such as cavity trays, which will be discussed later in this chapter. The polymeric materials provide a tough material that will not extrude under load, give good adherence to the mortar and have good durability even where bent to form gutters. Although naturally black, it is possible to obtain

Figure 19.6   DPC arrangements in cavity walls.

Thermal performance

With concerns for energy saving regarding winter heating in houses, the initial focus for thermal performance has been on insulation properties to resist the passage of heat. The focus is on the elemental U value or thermal transmittance of the wall as an individual element. Still air provides good insulation so the unventilated cavity not only gives the weatherproofing that is its main role but also contributes to the overall thermal resistance of the wall as long as it is at least 25 mm wide. This can be enhanced by providing a low emissive surface facing the cavity to limit radiation transfer. It is also possible to increase the resistance of the wall by careful specification of the block. Concrete blocks vary in their density. It has already been established that higher densities give greater strength and thermal capacity but lower insulation values. Selecting a low-density block can significantly affect the insulation value of the wall but still provides sufficient strength for a two-storey house. However, even with the cavity and the low-density blocks, the wall will not have a thermal transmittance or U value sufficient to meet either the current regulations or the expectations of wider environmental concerns.

It is possible to increase the thickness of the most effective material (in this case the block, as increasing the width of the cavity beyond 25 mm will have little effect) or introduce a layer of material specifically for its insulating properties. The most common solution is to include an insulating material in the cavity. Either partial fill (still leaving a cavity) or total fill solutions can be achieved with bats or boards and total fill solutions can be achieved with loose materials or in-situ processes installed after the walls are completed. Many combinations of block and insulation within the cavity can achieve elemental U values of 0.35 W/m^2 K and better with a 100 mm block and around 50 mm of insulation. Figure 19.7 shows some typical combinations of block and insulation bats and boards used as a partial cavity fill retaining the insulating contribution of the cavity.

It has been recognised that the wall itself is only one part of the construction through which heat is lost. Doors, windows, roof and floor elements all have different U values and it is possible to trade-off high insulation value in one element for another as long as the overall heat loss from the house will not be affected. This leads to the specification of target U values that can modify the required performance of the wall.

Although the performance for the wall was initially set in terms of resistance to the passage of heat with its focus on limiting fabric losses, the wider purpose of insulation is the conservation of energy. Walls and their insulation are only one aspect of energy saving for the building as a whole. Elemental U values are still important but are increasingly only part of the scheme to limit the energy use in a house. Aspects of

	Material	Conductivity (W/mK)
A	Mineral wool insulation bats	0.033
B	Rigid foam insulation boards	0.024
C	Lightweight aggregate concrete block density 1100 kg/m³	0.34
D	Aerated concrete block density 480 kg/m³	0.11

Insulation/block combinations	Elemental U value of wall (W/m² K)
A–D	0.34
B–C	0.35
B–D	0.29

Figure 19.7 Partial fill insulation U values.

ventilation rates, water heating, internal gains and solar gains are considered along with fabric losses in energy rating schemes that look at the energy consumption of the building as a whole. The external wall may now have performance requirements beyond insulation. The openings will influence ventilation rates and the thermal capacity of the fabric can be used to harness the heat gains to reduce the overall heating requirements for the house. As thermal capacity becomes significant so the position of the insulation in the wall has to be considered.

To understand why this becomes significant it is necessary to consider the wall not in steady state but to see the temperatures on either side of the wall changing. If the wall has a high thermal capacity and takes a longer time to heat up towards steady state so it will help to limit the effect of heat gains raising air temperature. Further, when air temperature starts to fall then the heat 'stored' in the wall will be given back to the air, delaying the need for additional heat inputs. A similar cycle can be used for summer cooling in larger commercial buildings.

The position of the insulation will also affect the condensation risk. The thermal behaviour of the wall for condensation risk, response times and thermal capacity issues are explained in Chapter 10.

The other major aspect of the wall resisting the passage of heat is the concern for thermal bridging. With masonry construction, these occur at openings and junctions with the floor and roof. These details are discussed under these sections in this and other chapters.

Size of the cavity

It is now possible to return to confirm the size of the cavity that was initially set at 50 mm wide. It was suggested that in order to act as a barrier to penetrating damp the cavity has to be wide enough to be kept clean and this will be influenced by both the inclusion of insulation and the production process. The cavity should be at least 25 mm if it is to achieve the thermal insulation potential of the still air providing good surface resistance on both sides of the cavity, so the 50 mm is satisfactory for the thermal function.

The influence of the production process can now also be identified. The process will involve bringing up the inner skin while protecting the cavity from mortar dropping as the work proceeds. If insulating boards or bats are to be installed in the cavity these are then placed and restrained against the inner skin (quilts will not be stiff enough to give a stable face inside the cavity). It is then possible to lay the external bricks. Two aspects about laying the external brickwork now need to be questioned. How easy is it to clean the mortar that squeezes out from between the bricks as they are tapped to level and how accurately can the width of the cavity be maintained? While it was easy to clean the mortar from the blocks forming the inner skin, the bricks are forming the cavity and the bricklayer has to clean the mortar with the trowel from inside the cavity. If the cavity is too small there is the danger that the mortar protrudes into the cavity and cannot be cleaned or falls down into the cavity where it will build up at the bottom of the wall, at window heads and even catch on wall ties, providing a damp path from external to internal skin. This will be aggravated by any narrowing of the cavity to keep the external wall in line, given the accuracy that was achievable in laying the internal block skin first. Experience has shown that it is only fair to the bricklayer to have a cavity around 50 mm in order to keep the cavity clean to achieve the insulation value and eliminate the risk of penetrating damp.

Many options for insulating the wall involve the cavity in different ways, some of which have been introduced in the section on thermal performance in this chapter. As an example, a wall could be specified with a cavity with partial fill insulation of 50 mm and a clear cavity of 50 mm, giving the overall distance between the inner and outer leaf of 100 mm. Given that the wall tie is often in compression acting as a strut then

some thought has to be given to buckling as the distance between the skins increases. Wall ties were originally designed for clear 50 mm cavities with no insulation. They have to be embedded in the masonry joint around 60 to 65 mm and have to be stiff enough to resist buckling across the free length in the cavity. Ties for 100 mm wide cavities have therefore to be longer and thicker so the correct type associated with overall cavity width has to be specified.

The wall so far

The variety of options for thermal insulation, its position and type of material, with its thickness in conjunction with the specification for the block and its influence on the cavity, have already been identified. Appearance and exposure conditions lead to a variety of performance needs that influence the choice of the external skin. Making a single choice for this case study is therefore not possible, but Figure 19.8 shows one option adopting the partial fill cavity insulation given as an example in the paragraph above. This wall has been chosen to achieve weathering, structural and thermal performance. As specified, it will provide a good level of sound insulation as it has both mass and a discontinuity with the cavity and the material chosen will give

Figure 19.8 Basic specification for a cavity wall.

good fire resistance. The basic structural and environmental functions will be fulfilled by this construction.

Internal and external finishes

The external appearance has been set by the choice of facing brick, although detailing and the use of more than one brick and the colour of the mortar or the use of renders can give design options. The shape of the front of the mortar joint is also sometimes used as a design feature, but it also has a significant effect on the weathering and durability of the wall. These alternative joint finishes are shown in Figure 19.8. The struck or weathered joint is a tooled joint (it has the surface smoothed after the mortar has started to cure) which promotes the runoff of water and protects the brick, while the recessed joint has the opposite effect, leaving the soaked edge of the brick vulnerable to frost attack. The struck or weathered joint takes time and skill and therefore the more straightforward, tooled bucket handle joint is often specified. A fourth joint finish, the flush joint produced as the work proceeds without tooling, does not have good resistance to rain penetration and may not produce a good edge on uneven or open-textured bricks.

Internally, concrete blocks can be specified for a fair-face for direct painting, but this is unlikely to be satisfactory for a house. The options will be between a wet process, probably lightweight gypsum plaster, and a dry plasterboard system.

The blockwork provides the background for the finish. The characteristics of the background introduced in Chapter 17 for the floor finish are the same for the walls. It is necessary for the background for the finishes to provide:

- Sufficient stable support
- Suitable fixing opportunities
- Compatible flatness tolerances
- Matching for future movement
- Avoiding deterioration mechanisms

If plaster is used, then the background characteristics of the blockwork will be satisfactory for the applications of two-coat work. The wall is clearly stable as it is part of the structure.

The density of the blocks is significant in that it determines the suction during plastering that makes the bond that provides the fixing between the block and the plaster and this will influence the specification of the plaster. The tolerances on the wall should be well within the capability of the first coat of plaster, normally about 12 mm thick to bring the surface to a flat plane. The final 'skim' coat of plaster will then provide a tough, smooth surface for decoration. Plaster shrinkage should not be sufficient to break the bond with the wall, although shrinkage cracks may appear at the top and bottom of the wall. If blocks are inert then there should be no interaction causing deterioration of either block or plaster. However, if the blocks are laid and finished before they have completely cured and continue to shrink, then cracks may appear in the plaster.

The plaster surface, while tough enough for most domestic situations, takes only limited-impact loads and is not suitable for external exposure. If either of these conditions may be experienced, renders should be used. Based on sand and cement with a plasticiser, traditionally lime, these finishes can be used externally even in severe exposure conditions with the appropriate specification. If renders are used externally then, for economy, a concrete block could replace the brick external skin. As with plasters, the density of the block chosen would determine the strength of the background. The first coat of render would have to be specified to match the strength of the block, with the final coat(s) determined by the exposure conditions. Finishes on external renders can vary from smooth trowled finishes that can be painted to thrown aggregate finishes using decorative aggregates to provide a weathering surface.

Dry finishing systems for internal walls will most likely be based on the use of plasterboard. As a background the blocks are not sufficiently flat to apply the boards directly to the wall. The wall is stable and can provide for a range of fixings. Two options are possible: to provide timber battens levelled to receive the board to be fixed with nails or screws, or to use 'dabs' of a plaster-like material to both fix and level the board to the wall. With both these systems the boards still have to be jointed. While square-edged plasterboard (fixed showing the grey face) with the joints reinforced with scrim can then be skimmed with plaster, the use of tapered edge boards allows the joint (and screw holes) to be finished without having to plaster the whole wall.

Interaction with other elements of construction

The interactions with the floor and the roof in both load transfer (roof/floor to wall) and the provision of stability (wall by roof/floor) have been identified. Services also interact with the wall and therefore may influence the specification and detailing. There may have to be some hiding of wires below the plaster in chases in the internal skin, but most pipes will be surface mounted and then boxed in if there is a wish to hide them. Radiators and other components will be fixed to the wall, demanding some consideration of the strength of the wall and the ability to identify suitable fixing devices to take the loads involved. This will also be true of the need of most occupiers to have shelves and cupboards fixed to the wall particularly in kitchens. A range of fixings is available for a variety of block walls and even the low-density blocks have the capacity, with the correct fixings, to take these loads, so the specification so far suggested will meet these performance requirements.

Life cycle considerations

The choice of the external material with concern for the exposure conditions to ensure durability has been discussed, as has the choice of the internal finish. The insulation and the internal skin should remain dry, so, in the absence of infestations in the cavity damaging the insulation, there should be little deterioration of these materials, although loose fills may compact with time. Maintenance requirements should be minimal if the external bricks are laid in cement mortar. Modification and adaptation will have to be undertaken with care, as the wall is a load-bearing structure. Any changes to the external wall may also have difficulty in matching bricks

at some time in the future, but internal finishes are easy to reinstate.

Consideration of the disposal of the masonry components at the end of their life indicates opportunities for recycling as fill, as the insulation should be easy to separate from the demolition waste. It is conceivable that the insulation may also be retrieved as it is only restrained in position, although it may have mortar from the blockwork stuck to the back that may be difficult to remove. The simple inclusion of a paper layer between the insulation and the block so that the insulation does not bond to mortar may be an inexpensive way to ensure the insulation could be reused. Metal wall ties could also be recycled although it is unlikely that they would be in sufficiently good condition to be reused.

Openings in the external wall

The wall considered so far does not provide access, light or ventilation, which will be required in an external wall. For these we need doors and windows. These, however, cannot take loads and will have less thermal, sound and fire resistance. They also create a safety and security risk and may limit privacy. This is normally resolved by creating an opening to maintain the integrity and performance of the wall while designing independent components for the doors and windows to complement the wall and ensure the function of the wall as a whole. To this must be added the need to provide fixing and jointing of the components into the opening. While the size of the opening will be dictated by the design requirements of the door and/or window, the basic detailing of the opening for the wall to the house is unlikely to vary.

The terminology used in openings is given in Figure 19.9. The detailing of the head, jamb and sill (threshold) will each need to be considered separately but are linked in the need to keep the depth of the reveal and soffit constant around the opening. The head will need to ensure that the loads are transferred to the jambs. The jambs will then need to be stabilised as they take the additional compressive forces. The sill has no

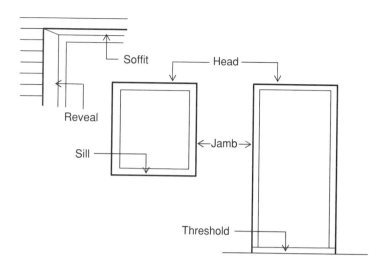

Figure 19.9   Openings in walls –
terminology.

structural requirements as it can be formed on
the wall below, but the weathering requirements
and the exposure conditions are more acute at
the sill.

All this has to be achieved with an accuracy
to ensure the doors and windows, now almost
certainly components made off site, can be fitted
and weathered into the opening. While some
packing may be tolerated below the sill the fit
between the jambs has to be accommodated in
the fixing and weathering details of component
to reveal.

Detailing the head

Carrying the load across the opening is fundamental to the detailing of the head. In the
past, arches have been used because the basic
material of the wall is weak in tension, so, with
the basic structural action of the arch being in
compression, the arch allowed the same wall
material to be used to span the opening. With
the greater range of materials now available the
lintel (a small beam with a bending action) will
almost certainly be used.

If the lintel is based on a rectangular cross
section carrying both skins as a simple beam,
the top of the lintel will collect water that can
run back towards the inside of the wall and the
front of the lintel may need a finish to improve
the external appearance and/or durability. The

rectangular section has been used in the past for
stone, timber and reinforced concrete lintels. All
these materials will create a thermal bridge if the
insulation is provided in the cavity, although
this would not be the case if internal or external
insulation were used.

While all these difficulties can be overcome,
developments in pressed steel and lightweight
concrete provide economic solutions that take
all these concerns into account. Two alternative
details of the head of an opening showing a steel
lintel and a lightweight concrete lintel are given
in Figure 19.10.

These solutions recognise that for the two-
storey wall the loads are on the internal skin and
that the need for strength to support the outer
skin is limited. In the lightweight reinforced
concrete solution the basic rectangular section
remains to support the loads from the internal
skin, although it is now shaped to form a cavity
tray to direct water away from the inner skin. The
gutter emerges at the steel shelf angle that supports the external skin, and weep holes formed
in some of the perpendicular joints between the
bricks help remove the water more efficiently.
It is now possible to bring the brickwork across
the opening to visually give continuity to the
external finish and protect the lintel.

The pressed steel solution gains its strength
not from a solid material but by shaping sheet
steel. The weathering details and appearance

Cavity gutter

Weep holes in brick joints

Weep holes in brick joints

Window or door frame

Window or door frame

Galvanised pressed steel lintel with insulation insert to prevent cold bridge

Aerated concrete (conductivity not exceeding 1.6 W/mK to prevent cold bridge)

Figure 19.10 Alternative head details.

remain the same as the composite concrete and steel shelf angle solution. The major advantage is the low weight of the lintel, an advantage in the production process. The hollow section can be filled with insulation to reduce cold bridging and some means of providing a background for the internal finish has to be incorporated under the lintel. The steel has to be protected from corrosion with a galvanising layer, but even this may be suspect over time in exposed conditions, particularly with the risk of damage during the production process. It is possible to specify polyester-coated lintels to provide extra protection.

One detail that cannot be seen from the cross sections in Figure 19.10 is that the cavity tray, forming the gutter from sheet DPC, should either extend beyond the lintel to minimise the risk of moisture running back off the end of the gutter and across the underside of the lintel, or use stop ends to the gutter to prevent the water running off the ends and wetting the cavity insulation.

Detailing the jamb and the sill

The detailing at the jamb is relatively simple and two options are shown in Figure 19.11.

Blockwork inner skin returned

Cavity closure with insulated core

Window or door frame

Wall ties at every block joint (225 mm)

Vertical DPC insulated to minimise cold bridge

Figure 19.11 Alternative jamb details.

The first shows the return of one skin to close the cavity and then providing a vertical DPC to eliminate damp penetration across the jamb. Which skin is returned will determine how deep the reveal is, as the frame of the window or the door will be located to cover the DPC, ensuring no moisture reaches the internal finishes or the back of the frame if this is against the external skin. Returning the block (detail shown) will reduce the thermal bridging effect but will leave the frame fairly close to the external face. The reveal will be small. This means that the sill on the frame can be detailed to hang over the wall without the need for a sub sill but brings the joint between the frame and the opening into a more exposed position. In order to give the joint better protection, the outer skin can be returned to gain a deeper reveal. A sub sill will now be required and the brick return creates more of a cold bridge, although insulated DPC can help reduce the effect. The second option is to use a cavity closure. This hollow plastic box section can be filled with insulation to reduce the cold bridge, act as a DPC and in addition can provide a fixing for the frame.

Structurally the edge of the wall has to take the load from the lintel but has lost the stabilising effect of the bonding and this makes the jamb subject to buckling. The return skin or the cavity closure provides support if the skins buckle towards each other, but provides no resistance to them moving apart. To provide this resistance, wall ties are placed in every block joint adjacent to the returned skin. The cavity closure can incorporate ties that attach to the box section.

The detailing of the sill, with and without a sub sill, is shown in Figure 19.12. The major concern is to shed the water that runs off the impermeable window or door surface away from the wall as effectively as possible. The top surface of the sill should be sloping, the lip should overhang the wall and the underside of the lip must be provided with a drip. This amount of water around the sill raises the concerns of durability so the material of the sill needs to be chosen with care. Moisture penetration is an additional concern, particularly at the ends of the sill at the reveal and under the frame where wind-blown rain will be held and may even be under pressure rather than running off down the slope provided on the sill. If the cavity is closed in the sill detail then a DPC will be required to protect the internal finishes, although bedding the sill on a DPC is common practice to protect a timber frame sitting on the external skin. Care has to be taken to ensure that the edge of the DPC is not laid forward of the line of the front of the wall or the drip may be breached and water drawn back under the sill. Cavity closure sections can also be used in the sill detail.

Doors and windows – the components in the openings

It is not the intention to consider the details of the components themselves. This has become a

Figure 19.12   Alternative sill details.

specialist area of expertise, where the manufacturing of sections and the factory assembly of the components have to be understood as well as the component design aspects such as strength, weathering, safety and security. All these are beyond the scope of this text. However, the design of doors and particularly windows with the joint and fixings into the openings needs to be thought through, particularly as concerns for sustainable development are leading to a re-evaluation of energy saving by considering the behaviour of the house as a whole. This includes a consideration of the orientation of the façade, the area or percentage of glazing and the shape of the windows, as well as the specification of the materials involved. While this has a significant impact on the look of the house and the views experienced by the occupants it also has an impact on the energy efficiency of the house.

Given the predominant energy used in houses during winter heating conditions, any contribution that can be made to reducing heat loss or increasing gains needs to be considered. Windows have the potential to do both. Solar gain from south-facing facades can be usefully used for many days even in winter, where direct radiation from the sun will pass through the glass and heat objects in the room, adding to the heat gains. However, the glass is poor at retaining the heat from the air in the room, increasing heat loss. Heat loss can be considerably reduced not only with double (or triple) glazing but also by the choice of glass with coatings or gas-filled cavities which can reduce U values of double glazing almost to that of triple without the expense of the more complex frames and opening casements.

Using glazing for heat gains is only effective on facades facing in a southerly direction. In these facades larger areas of glazing can provide a net gain, but this is not true for other facades; here it is better to limit the area as heat gains are not possible and the wall construction can provide much higher U values to reduce heat loss. However, double glazing with coated glass can still make a significant contribution to reducing heat loss on these facades.

These choices to reduce winter energy use have consequences for both lighting in the rooms and heat gains during the summer. Light from the south can be very bright and cause glare, so larger windows may create discomfort or even disability during the summer, requiring the use of blinds or curtains. Similarly, heat gains will be greater, possibly requiring shading. Conversely, while north light is good quality and more even, the smaller windows may make rooms dim during the winter, requiring artificial lighting and adding to energy use. Heat gain is unlikely to be a major problem from the north.

Currently, the number of days when summer heat gain or poor lighting are likely to be of concern in the UK is limited as for many households daytime occupation is limited (unlike commercial property where this is a major problem, particularly with high heat gains from equipment and lighting). For the days when conditions are uncomfortable, ventilation by opening windows to gain cross flow is the main mechanism offered to householders to control the internal temperature. If either climate changes or occupation patterns vary then this may have to be reconsidered, and designs including light-coloured walls, screening and even verandas used for houses in hot climates might be considered. Perhaps high thermal capacity construction with night-time purging being used in commercial property in temperate climates might be adopted for house construction.

Having considered shape and size along with any additional components such as blinds or shading, the choice of materials has to be considered. Glass has been discussed but not the choice of material for the frame. For domestic buildings, timber (soft and hardwood) or plastic (uPVC) are the main choices. Aluminium is also specified, but it has a higher capital cost than timber or plastic. Steel has in the past proved to have a high maintenance cost to control corrosion. However, galvanised, polyester coatings and stainless steel overcome these problems of the past but are not often considered for housing given the viable options of timber, plastic and even aluminium, which can give slender sections and be produced with coloured anodising if this is required for appearance. The choice between these frame materials will include an analysis of the need for maintenance, potential life, ease of replacement and potential recycling and disposal options. Timber (particularly hardwoods) can have an impact on habitat and global climate, while plastics have pollution and energy considerations associated with their manufacture.

Having chosen the size, shape and material for the windows and doors and had them manufactured to give strength, be weather-tight and secure, it is necessary to consider how they are to be fixed and weathered in the opening.

Windows and doors can be built in as the work proceeds, being fixed back to the brickwork with frame cramps or secured to the cavity closure. They can also be installed after the brickwork is complete by screwing through the jambs. If they are fixed after the brickwork is complete the bricklayer has lost any direct control on tolerances to maintain the gap between frame and brickwork where the appearance and weathering functions have to be achieved. This will have to be replaced either by careful measurement of opening size, square and plumb or with the use of temporary frame forms. Packing will then have to be used between frame and brickwork to ensure the jambs of the frame are not distorted as the window is secured.

However the frame is fixed, weatherproofing is normally achieved with gunned mastic, although consideration should be given to placing the window towards the back of the reveal to give greater protection, particularly in exposed conditions.

The panel frame option

The construction analysed so far has been based on an initial suggestion of masonry skins to the wall. These have proved to be capable, providing the required performance and being able to be detailed around openings and at the connections to other elements of construction. Given the context of the case study described in Chapter 16, a viable alternative to the masonry internal skin that might be considered is a wall based on the framed panel.

The panels can be formed on site but are more commonly formed in a factory and delivered to site as flat panels. Some developments have built the panels into rooms in the factory that are delivered to site as 'volumetric' units. The basic form of the loadbearing panel in timber was shown in Figure 19.3. Similar construction can be achieved with light steel sections. The following text only considers timber, but the process of analysis could be applied to the light steel solution.

Timber frame construction

Suggesting that the inner skin is formed from timber frame panels rather than a concrete block affects nearly all the aspects of specification and detailing of the wall. It has such a fundamental effect that it is referred to as 'timber frame construction' even though it is only the inner skin that has been changed in principle.

The origins of these fundamental effects can, for the most part, be attributed to the change already identified:

- The basic form (solid masonry to open panel)
- The material (concrete to timber)
- The production process (site operations to factory prefabrication)

The basic form

The component layers of a typical timber frame inner skin are shown in Figure 19.13. The basic form of open studwork with a sheathing board on one side allows the insulation to be introduced within the thickness of the skin. The temperature profile will now show the condensation risk on the cold side of the insulation, leading to the need for a vapour control layer behind the internal finish. This vapour control layer is likely to be a polythene sheet where, even with care on lapping and sealing edges, it can only be considered to be a control and not a vapour barrier. To ensure the control is effective it is

Plasterboard internal lining

Polythene vapour control

Mineral wool thermal insulation

Breather paper

Structural sheathing board

Bottom rail

Studs at 400 or 600 mm centres

Figure 19.13 Inner skin of timber frame cavity wall.

necessary to ensure what is known as the five times rule. The vapour permeability on the warm side of the insulation (including the vapour control layer) should be at least five times that of the rest of the construction on the cold side. It is therefore necessary to choose the materials on the cold side with care to ensure they have high vapour permeability if condensation is to be controlled. It is usual to specify insulation materials that can breathe, such as open fibre mineral wool bats or quilts.

For structural purposes in a typical panel for a two-storey house the studs need only be around 89 × 38 mm at 600 mm centres, although this limits the thickness of insulation if it is only to be provided between the studs. For increased strength and/or insulation 140 × 38 mm studs are common, but for increased insulation 180 × 47 mm might be considered.

The open nature of the studwork also affects the possibilities of providing fixings. Components transferring load, such as wall ties, must be fixed to studs and therefore have to be at centres to match the studs. Heavy items inside the building need to be identified because if they cannot be fixed to studs they may need to have noggins incorporated.

The open framework also dictates the finish. The studs do not provide a continuous background but do form a plane with good flatness tolerance, the timber gives easy fixings and a plasterboard finish is most appropriate. Screwing the plasterboard to the studs and tapered edge boards to save a skim coat is the most likely specification.

Influences of the material

Perhaps as fundamental as the open form of the panel is the change of materials. The change of materials brings with it changes in properties. Those most significant to the performance of the internal skin of the wall are movement, fire and some aspects of durability.

The external clay brick skin will tend to expand, while both concrete and timber will shrink. This differential movement in the masonry wall is not significant and wall ties and lintels that span both skins have sufficient flexibility to accommodate this movement over two storeys. This is not the case with the timber/brickwork combination. Wall ties have to be flexible or designed in some way to accommodate vertical but not permit horizontal movement. A clear 12 mm gap at the top of the brick wall to the roof structure resting on the inner skin is needed for a two-storey house (assuming the inner skin is resting on a concrete ground floor). Window and door frames should be fixed to the timber internal skin and then gaps have to be left between the brickwork and the underside of the sill and the head of windows. These gaps will have to have a compressible filler and flexible sealant to weather the joint.

The fire resistance of the timber will be less than the block skin. While the brick maintains the fire resistance of the wall as a whole, the fire may reach the cavity and this provides a clear path particularly to spread the fire to other rooms and the roof. This requires the provision of cavity battens, the regulations and recommendations for which vary depending on location in the UK. They are required horizontally at the top of the wall to protect the roof and may be required at each floor level. They are normally provided around all openings and are required vertically at party walls. The cavity batten has to be in contact with both the cladding and the sheathing in order to act as a barrier to the spread of fire in the cavity, but this will bridge the cavity, providing a potential path for damp penetration. These cavity battens may be ridged (38 mm preserved timber battens on DPC) or flexible (50 mm wire-reinforced or polythene-sleeved mineral wool) with polythene DPC tray over all horizontal battens.

Consideration of the deterioration of the timber affects the detailing. Timber has to be maintained at moisture contents below 20% to inhibit decay. During production, finishes should not be completed if they might seal in timbers with values above this limit. Detailing should be devised to keep moisture below this level during the life of the building. Where there is a risk of this being exceeded then durable species or preservative should be used. As the timber likely to be chosen is unlikely to be a recognised durable species, preservative may be specified. With careful detailing it should not be necessary to preserve all timbers in the house. However, some timbers are difficult to protect with detailing alone. Preservative treatment is essential for sole plates, timber against DPCs and cavity battens, although for insurance reasons the preservation of all the external wall frame timbers may be recommended.

While for production reasons the cavity will still have to be 50 mm, if the cladding is brickwork it will have to be drained and ventilated. The buildup of moisture in the cavity is a threat to the timber so a flow of air helps keep levels of humidity low. In addition, a breather paper protects the panel during construction until the brickwork is completed. The breather paper is waterproof to minimise the risk of moisture being absorbed by the timber but has good vapour permeability to ensure the five times rule is maintained.

The production process

The production process is determined by decisions about how large the panels can be to transport to site and handle into position. The basic floor and roof construction of timber joists and trussed rafters is still applicable, but it is now sensible to consider some level of prefabrication or at least preparation to co-ordinate all these timber elements with the walls. In the UK, single-storey height panels are fixed above DPC level where floor cassettes are fixed at first floor level bearing on the top of the panels. From this new platform the first floor wall panels are erected where the roof trusses and bracing are fixed to complete the structure. It is also possible to prefabricate the internal walls and erect these at the same time. This storey by storey height process with floors resting on the top of the panels is known as a platform frame; a multi-storey height wall with floors hung off the wall known as a balloon frame has been developed but is not often used in the UK.

It is possible to conceive of a wide range of levels of prefabrication for the platform frame system for all the elements of the house. In the UK most wall panels are delivered as 'open' panels. The basic stud and sheathing board panel often with breather paper is factory produced. The panels can be easily nailed together and the breather paper laps arranged on site. It is possible to think through the prefabrication of 'closed' panels where insulation, services, vapour control and wall linings are fixed in the factory. This makes site process connecting panels more complex and the need for site protection more acute.

Whatever the level of prefabrication, it will require some co-ordinating dimensional discipline to take full advantage of the factory process and this will have to be applied to the design of the house and the brickwork outer skin. It will be necessary to establish a grid on plan and decide on some vertical heights. This idea of a dimensional reference framework was discussed in Chapter 4.

All of these structural elements (walls, floors and roof) are erected in one set of operations, with the outer skin or cladding being added later. This is a shorter site operation compared with completing the structure with masonry walls. Scaffolding has to be erected to eaves level before the erection procedure is started and a crane will be required throughout the procedure.

This production sequence means that the brick skin has to be complete with the whole of the internal skin in place. The brickwork is now highly dependent on the accuracy with which the panels have been erected. Normal brickwork tolerances are ±10 mm in 5 metres length and ±20 mm in 6 metres height. The timber panels should be erected within these limits, in which case the bricklayer can maintain the width of the cavity to ensure that wall ties, cavity battens and any lintels attached to the panel should all be installed within their tolerances. Erecting the timber frame panels within tolerances to the dimensional discipline grid is also important to allow the brick bonding to be set out without excessive cutting of bricks or variations in the thickness of the joints.

Alternative cladding options

Render is possible with a timber frame, but it is not likely to be applied to a block as it was in the masonry wall but to metal lathing fixed to pre-served battens fixed directly to the studs. If the lathing is unbacked (render can squeeze through to behind the lathing) the cavity has to be 50 mm to keep it clean, but with backed lathing systems the cavity can be reduced to 25 mm wide.

Other claddings such as tile hanging and board cladding can be adopted, again fixed to preservative-treated battens. With tile hanging the battens are horizontal so there is no vertical cavity, but with lap jointing each tile can drain independently. With boarding the joints cannot drain and a 10 mm wide cavity should be formed. In this case the insulation value of these cavities is not very great, but they will fulfil the waterproofing function.

Care has to be taken in specifying these alternatives. The bricks provide the major part of the fire and sound resistance to the whole wall. Changing the cladding may leave the building unprotected in these areas. Some cladding may be combustible and have poor surface spread of flame. Sound insulation is dependent on other elements such as windows and doors, but the contribution of the wall has to be considered.

Party (separating) walls

Unlike external walls these walls have no weathering function and will have no openings, making their construction generally more straightforward. However, they may be load-bearing (including racking forces from wind loading) and will need greater performance in both fire and sound resistance, and this will determine most of the construction specification and detailing. The basic construction options are shown in Figure 19.14.

In masonry construction both fire and sound insulation can be better achieved with dense concrete rather than the lower-density blocks chosen for their thermal properties for the external wall. The cavity is not necessary for weatherproofing but does provide sound reduction, so both solid and cavity construction can be used. For fire resistance the wall has to extend into the roof space and be fire stopped up to the underside of the tiles.

For a solid party wall the total mass should be at least 415 kg/m^2 and this can be achieved with a 215 mm wide dense block or the wall can be formed by laying thinner blocks flat to form a 215 mm thick wall. Both of these can then be finished on both sides with light-weight plaster or a single sheet of 12.5 mm plasterboard as discussed for finishes to external walls.

If a cavity wall is used then two 100 mm skins with a 75 mm cavity (with wall ties) can be formed in lighter blocks with densities around 1500 kg/m^3, again finished with lightweight plaster or plasterboard both sides. In both these

215

Dense block at
least 1850 kg/m^3

19 mm plus 12.5 mm
plasterboard sheets
both sides

250

Plaster or plaster
board finish both
sides

275

Light metal ties

Wall ties in
75 mm
cavity

89 x 38 timber studs

50 mm mineral
wool insulation

100 thick blocks density
at least 1500 kg/m^3

Figure 19.14   Alternative party wall options.

specifications it is important to have the mortar joints fully filled to ensure the full performance is achieved.

Party walls can be constructed in timber frame construction as cavity walls. In this case it is difficult to achieve mass and fire resistance directly from the structure of the timber frame panel. The wall has to be constructed of two stud frames with a cavity between. The finish will have to be two sheets of plasterboard, one 19 mm and one 12.5 mm (staggered joints) on either side of the wall, and with at least 50 mm (80 mm recommended) of mineral wool quilt with a density not less than 10 kg/m^3 in the panel(s) facing the cavity. At floor level a fire cavity barrier should be provided and a row of lightweight metal cavity ties is required to stabilise the two skins for each storey. Fire barriers are also required vertically where the party wall meets the external walls while maintaining the sound discontinuity to avoid flanking paths. This construction relies on discontinuity and absorption in the cavity for its sound insulation, with the plasterboard providing some mass and the fire resistance between the dwellings.

This construction of the party wall has no sheathing board on the timber frame, which was a major component of the structural action of the panel particularly against racking. The plaster-

board is now fulfilling this function and may not be sufficient. If additional stiffness is required this can be gained with diagonal bracing inside the cavity or a sheathing board may have to be reintroduced into the construction.

Partition walls

Like external and party walls, partitions can be constructed in either masonry or framed panels. The basic construction is the same as the walls shown in Figures 19.1 (similar to block internal skin) and 19.3 (timber or steel stud), as are the options for finishes with plaster or plasterboard, as appropriate. Proprietary partitions where panels are delivered to site finished both sides ready for decoration are available but are generally more expensive.

Partitions are likely to have lower performance demands than either external or party walls. Walls immediately below the roof will not have to carry loads if the roof is constructed with trussed rafters. Walls below floors, however, may have to carry loads from the joists. There may be a need for internal walls to act as buttresses against overturning from wind loads, in which case they will need to be bonded or otherwise securely tied to the external wall over their entire height. Otherwise the partition can just be built up to the inner skin. When partitions are built against timber framed party walls both sheets of plasterboard on the party wall have to be fixed first to ensure no weakness in either sound or fire performance of the party wall.

Given no structural role, the choice of partitions within a dwelling will be dominated by the functions associated with appearance and privacy. They do have a nominal half-hour fire resistance requirement, but this should be satisfied with masonry with plaster or plasterboard, or a framed stud partition with 12.5 mm plasterboard on either side for the finish.

Appearance demands are normally limited to plain surfaces that can be achieved with plaster or plasterboards and the basic construction will provide a complete visual and light barrier. Sound may become the limiting performance for the choice of partitions. There is a requirement for construction of partitions between bedrooms and between rooms with a water closet and any other room to achieve an R_w value of 40 dB. This cannot be achieved in a stud wall without including a sound-absorbing quilt at least 25 mm thick between 75×38 mm studs. Block walls to achieve this sound rating will have to have a mass of $120 \, kg/m^2$ or more.

The thickness of partitions in a house is likely to be between 100 and 125 mm. This should ensure stability for normal domestic floor to ceiling heights with head restraint for lightweight framed panels, as discussed in Chapter 17. If it is required to fix items to the block partition this should not be a problem, but for the frame studwork the need for noggins (marked on the outside of the boards) for fixings for heavy items will be necessary. The interaction of partitions with other elements, particularly services, is the same as discussed for the external wall in the section above. Indeed, the finishes may well be chosen to be similar for all internal surfaces of the rooms whether they are to external, separating or partition walls.

Summary

(1) The three wall types discussed, external, separating and partition, each have a different set of functions to perform.

(2) Each of these wall types can be built either as masonry, panel or monolithic, which in practice are normally constructed in brick and block, timber or steel frame, or as in-situ concrete, respectively, although other options are available.

(3) The cavity wall, chosen for its weatherproofing action, is the most likely option for the external walls for the house in the case study, specified as either masonry or timber frame construction.

(4) The cavity that provides the weatherproofing is formed between an outer skin that determines the appearance and an inner skin that normally takes the loads. The cavity and the choice of lightweight blocks for the inner skin provide some thermal insulation, but this has to be augmented with a layer of insulation to achieve what would now be considered as a reasonable U value.

(5) To provide all the functions of an external wall, windows and doors are required. To accommodate these components, openings are formed in the wall, which have to maintain the performance of the basic wall. The components are then fixed and weathered into the openings.

(6) The components are now normally designed and manufactured by specialists and require only detailing to fix and weather the units into the openings. However, the size of the opening, the performance and operation of the components and the materials from which they are made must be chosen as part of the overall environmental design for the house.

(7) Timber frame construction in based on creating the internal skin of the cavity wall in a timber framed panel rather than blockwork. This change to panel construction in a different material together with the changes this brings about in the production process have a profound effect on the rest of the construction details, requiring a re-evaluation of most aspects of the structure and envelope of the house.

(8) Separating or party walls between dwellings have both fire and sound performance requirements that dictate most of the choice of materials and detailing.

(9) The options for internal partitions are also based on either masonry or stud work. They may be loadbearing downstairs and may have sound transmission requirements, particularly between bedrooms. The requirements for the integration of services and the finishes for partitions are similar to those for external and separating walls.

20 Foundations

This chapter considers the construction up to damp-proof course (DPC) but not including the ground floor. It introduces the influence of the soil and the site on both design and production for a range of foundations that might be considered for the house in the case study.

Basic function

It is the function of the foundations to maintain the integrity of the structure and fabric of the construction above ground. In this text the foundations will be considered to be all works up to the DPC for the wall but not including any construction associated with the ground floor that is taking loads from the activities on that floor (covered in Chapter 17). The foundations, therefore, have no environmental functions. They are structural and have some small appearance considerations where the wall shows above ground up to the DPC. Analysis of the life cycle is necessary as maintenance targets should be zero and the foundation will affect the possibilities in redevelopment of the site when the building is modified or demolished.

Structural integrity

Given that the primary structural role is to maintain the integrity of the building, two things need to be known about the structure and fabric above ground:

(1) The distribution, magnitude and direction/ point of application of the loads
(2) The stiffness of the construction above ground

For the house, all the loads come from walls, arriving at the foundations as uniformly distributed strip loads. Each wall will have a different load per metre run depending on which walls take the roof and floors. Wall loading for housing also depends on self weight of the wall, which can be a significant part of the load

particularly if constructed in masonry and with few or small openings. While the loads from external and party walls vary, the difference is unlikely to be great and therefore it is usual to specify all foundations to these walls based on the greatest loading. This will theoretically lead to different settlements under each wall, but in practice, on reasonably good bearing soil, these variations in loading do not lead to significantly different settlements that would cause distress to the superstructure. However, the loads on internal partitions will be considerably less and the wall will be thinner so smaller foundations would be chosen.

House construction will produce vertical loads at the foundations and, although theoretically most of the imposed load will be from the internal skin, the load will be taken to be acting through the centreline of the whole wall.

The stiffness of the building, its structure and fabric, is a measure of how much differential settlement it can take before it suffers any excessive distortion or cracking that would lead to a failure of performance. Initially, distortion or cracking may only affect appearance, but as they become more pronounced then weathering (serviceability) and structural stability may be threatened. All buildings distort and crack to some extent and much detailing is designed to accommodate these movements. It is this detailing and the choice of materials that determine how much movement can be accommodated without affecting performance. In house construction a major consideration is the stiffness of the block/brickwork. As the stiffness of the structures increases, critical strains (the movement that develops ultimate stress and cracking) will be reached with smaller differential settlements. The use of cement mortars for strength makes the wall relatively stiff and less able to take differential settlement than walls in weaker, less rigid lime mortars. The nature of the connections of the floors and roof to the wall being pinned allows some rotation but they are susceptible to lateral movement which may involve a reduction of the bearing.

Soil as part of the structure

The soil under the foundation will be stressed and is therefore part of the structural system of the building. The foundation of the building must act to transfer the load in such a way that neither the foundation nor the soil is overstressed. The analysis of the behaviour of the soil under load is given in Chapter 11. It identifies that the compressibility of the soil below the foundation to a depth of around 1.5 times the width of the foundation will determine the settlement, while the soil to the side of the foundation will be involved in a collapse failure dependent on the soil's shear strength. All this is dependent on the absence of instabilities in the mass of soil outside this area of stressed soil caused by volume change from moisture movements or soil collapse from slope instabilities or voids caused by natural or mining activity.

This analysis identifies what needs to be known about the soil and the extent of investigation that is required to establish good information on which to base the design of the foundation.

The basic soil properties of the shear strength to determine the ultimate bearing capacity and the compressibility (including time-dependent behaviour) have to be known. In addition to these properties that are related to the capacity of the soil to resist the loads, it is important to know about the soil's volume change with changes in moisture content.

It is characteristic of clays (and the more cohesive silts) that changes in moisture content are accompanied by changes in volume. Removing water from these soils below foundations leads to shrinkage which would cause settlement additional to that expected from the application of loads. Increased water content in these soils causes swelling. If additional water becomes available, the suction forces in the very fine structure of cohesive soils is great enough to overcome the weight of the building and the ground will heave, lifting the building up. Once it is established that the soil on the site is subject to these volume changes, it is necessary to ask to what depth these moisture changes can be expected.

There are two major causes of moisture change: seasonal changes and the vegetation (mainly trees) around the building. In the yearly cycle between summer and winter the soil near the surface will dry out (desiccate) from the sun and take up moisture again with the increased incidence of rain. In the UK these seasonal changes are variable, but over the life of the building some dry summers and wet winters will be experienced. Past records have shown that these are unlikely to affect the soil below about 900 mm so any foundation below 1000 mm should not experience this seasonal volume change in the life of the building. Vegetation, however, can extend its influence below this level. The effect is very dependent on the species of the tree, its thirst for water and its mature height, and how far it is from the building. Mature trees will draw water from deeper and wider in dry summers, leading to shrinkage and settlements. If mature trees are removed they will cease to take water and the soil will return to a higher water content, causing swelling and heave. The depth of soil affected has to be assessed for each site and incidence of trees but can be as deep as 3500 mm or more, much greater than seasonal influences.

There is one other volume change associated with the presence of moisture, but this is not from change in moisture content but in the freezing of any moisture in the soil. Soils susceptible to movement under these conditions include silts, chalk, fine silty sands and some lean clay. This can cause heave if freezing takes place below the foundation. For housing this is a seasonal change associated with very cold winters with prolonged sub-zero temperatures. Again, the winters in the UK are very variable in this respect, but during the life of the building, past records have shown that in areas adjacent to occupied buildings depths of up to 450 mm have been affected, so foundations below 500 mm should not be affected. For specialist installations such as cold stores, greater depths may have to be considered.

Site investigation

It is now possible to identify what needs to be known about the soil to a depth that includes the zone of stress and moisture changes if the soil is expected to be clay. It is also necessary to be clear how these might vary under the building, including the possible shear zone around the building. If the position of the building is not known then the whole site must be considered in the investigation. Beyond this, other information is necessary to consider the possibilities of subsidence or other ground collapse which might be a potential failure to be allowed for in the design.

In establishing a site investigation, it is also necessary to gather information important to the production process. Work in the ground carries with it risks directly related to the ground conditions. These are not only technical risks but also health and safety and financial risks which tend to increase the cost of works below ground. Selecting foundation solutions that minimise these risks is part of the process of choice. The soil properties identified for the design will give information on the risk of collapse of exposed excavations and the need for temporary support, but additional information, such as ground (standing) water and access possibilities, needs to be established.

Information for any environmental analysis must also be collected at this time. This might be concerned with habitat and the local eco-systems on a green-field site, but needs also to be concerned with previous land use and the possibility of contamination on brown-field sites. Care must be taken in making this distinction between green- and brown-field sites. Some

areas of the UK now considered to be wild and unspoilt might have had an industrial past from Roman and even Medieval times. Previous development may indicate the existence of old foundations or other construction below ground or contamination. It might also reveal antiquities or other archaeological remains.

Not all of this information can be gathered in the same way. This leads to most investigations being based on some or all of three processes:

- Desk study
- Walk-over study
- Ground investigation

The desk study uses existing published information. This might include climatic data, but it is most likely to be based on geological and topographical maps and air photographs. This study will yield information to assess the wider possibilities of subsidence from both geological and man-made landforms that might make the site unstable. This would therefore include records of old mine workings or maps that would show previous land use. The names of features or settlements are often clues to previous use. The desk study would also include surveys of the utilities – water, electricity, sewers, etc. – which can be obtained from the appropriate authorities.

The desk study would indicate what might be found on the site and will help to inform the other two phases of on-site investigations. Along with any local knowledge, it will give some idea of what land features and vegetation would be visible from above ground and what soil type, previous development and contamination might be below.

What is considered as 'contamination' might come from two concerns: the adverse effect it may have on the health of the occupiers (and constructors) of the building and the possible aggressive effects it might have on the materials to be chosen for the construction. This will determine what might be considered as sufficient levels of concentrations in the soil or ground water to be considered 'contaminated'. For these reasons, contamination surveys on brown-field sites are often undertaken independently of the engineering soil survey.

The walk-over survey gathers information from above ground. This would include the topographical and vegetational features and the conditions of adjacent development. It may be possible to just mark these on an existing site plan plus a photographic record, but it may show the need for a more accurate land survey if the positions of features or the exact levels of the land were important to the proposed building, its form and position.

The scope of the ground investigation will depend on what the desk and walk-over study identified and the size and possible construction of the building. For house foundations, simple trial pits will probably be sufficient, with the soils being identified with the use of soil classifications from visual inspection. If the site provides good bearing conditions then simple strip foundations can be designed to satisfy the Building Regulations from a table in the Approved Documents where only soil type is needed to select a foundation width. Simple visual classification will also be sufficient to determine the presence of the shrinkable clays important to specifying sufficient depth to ensure volume stability. Trial pits will give an indication of standing water and will be sufficient to take water and disturbed soil samples for chemical analyses, such as sulphates, if these are suspected. Where visual identification of soil type is not sufficient, then in-situ tests and the removal of undisturbed samples for laboratory tests, possibly from boreholes, may be required.

It is important to consider a fourth phase that takes place during construction where the conditions anticipated by the site investigation and used for the design need to be continually compared to the information revealed by excavations for the building. This will also lead to inspections prior to significant events, such as pouring concrete for the foundations themselves.

Foundation design

Whether the foundations are chosen from tables or are the subject of a structural design, four conditions must be satisfied:

(1) The load has to be distributed over a sufficient area to limit settlement and avoid collapse.
(2) The centre of gravity of the load has to be over the centre of the area of the foundation.
(3) The foundation has to be stiff enough to press evenly on the ground over the whole area of the foundation.
(4) The foundation has to be founded at a depth below ground to be resting on a sufficiently stable soil mass.

For the house under consideration in the case study, the loads are relatively low and many sites in the UK have soil with a good bearing capacity and low compressibility, although it is more than likely to be shrinkable clay in England, particularly in the south and east. If there are no major previous developments or geological or typographical difficulties, the simple strip footing will be the economic choice.

The strip foundation

The basic form of the strip foundations is shown in Figure 20.1. The external cavity wall rests on a DPC to stop rising damp. The DPC has to be at least 150 mm above the ground to avoid soaking of the wall from splashing back from the ground. The first part of the wall below the DPC has to remain as a cavity wall to ensure the cavity remains clear and can drain below the DPC. The external skin has to remain brickwork for at least two or three courses below ground to maintain the appearance, but this assumes the brick will remain durable in this zone where saturation and freezing are more likely. Below this, the external skin can be constructed in blockwork. Nearly all blocks are suitable for construction below ground so a change from the insulating block used in the inner skin above the DPC can be made as long as this does not create a cold bridge at the edge of the slab.

From 150 mm below the lowest DPC, the cavity has to be filled, as the wall will be subject to pressure from the soil tending to collapse the

Figure 20.1 Plain concrete strip foundation.

two skins together. It is possible to use a single 'trench' block the total width of the wall to avoid having to build the two skins and then fill the cavity.

The bottom of the wall then needs to be stiffened and probably made wider to transfer its load evenly to the soil. If the width does not need to be too much wider than the wall then a strip of plain (unreinforced) concrete can be used. This is the case for many sites in the UK where foundation widths between 300 to 600 mm will be sufficient. This concrete will not have to have great strength but will need to be durable and therefore the specification needs to include the minimum cement content (and type if sulphate resisting is required), as this will be governed by durability considerations.

Figure 20.2 shows how the design principles given above are observed. First the load and type of soil will determine a width to provide sufficient area. The wall must then be detailed centrally on the strip. To provide a stiff base to the wall in its length, the footing must be at least 150 mm thick. However, to ensure the load is transferred across the whole width of the strip, the thickness must be equal to the projection. This is related to the choice of the material as plain concrete. The strip will transfer the load to an area under the footing as long as the material remains in compression inside the 45° punching shear line.

This requirement for the thickness to equal the projection will give thicker foundations as the width increases. As the thickness increases it may be worth considering the possible change to the material whose structural behaviour is not dictated by the 45° punching shear failure in plain concrete. This can be achieved by introducing steel reinforcement into the concrete to provide a 'wide' strip foundation, as illustrated in Figure 20.3. The foundation can now sustain a bending action, ensuring the load is spread across the full width of the strip. The introduction of reinforcement requires the specification of cover. This is the thickness of concrete required to gain a bond between concrete and steel to develop a composite action in bending and to protect the steel from corrosion (above ground this may also be determined by fire requirements). For foundations this is determined by corrosion protection and will be 40 to 50 mm, depending on conditions. To ensure this cover is achieved in the production process, a concrete blinding is required and the reinforcement will be supported on spacers standing on the blinding. In this way the reinforcement will not be 'trodden' into the ground during the concreting operations.

Returning to Figure 20.2, the last parameter is the founding depth below ground to ensure the foundation is resting on a sufficiently stable soil mass. Given that the top of the strip has to

Figure 20.2   Design considerations for plain concrete strip foundation.

Figure 20.3   Reinforced concrete 'wide' strip foundation.

be at least two or three brick courses below ground, it is unlikely that the depth of the foundation (to the underside of the strip) will be less than 300 mm, so it should be below the level of the topsoil (normally around 100 to 150 mm in the UK). However, this is not below the frost line, requiring depths of 500 mm, or the influence of seasonal moisture variations on clay at 1000 mm, so these may determine the depth on soils susceptible to movement under these conditions. Deeper foundation may have to be considered if there is vegetation on shrinkable clays or on any site where geological, old industrial or topological features have to be taken into account.

The simplest topological feature is a sloping site. Even if it is not sufficient to make the site unstable, it will affect the foundation depth. The specification of depth below ground will have to be for the lowest part of the site, so, if the whole strip is laid to the same level, the depth of the foundation will increase at the rate that the level of the ground rises.

As the depth increases the production risks increase. There is a greater need for working space and temporary earthwork support. This will need larger machines and require the removal of more soil. All of these increase the health, safety and environmental risks and hence both cost and time will increase. These concerns

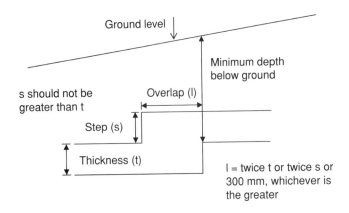

s should not be greater than t

Overlap (l)

Step (s)

Thickness (t)

Ground level

Minimum depth below ground

l = twice t or twice s or 300 mm, whichever is the greater

Figure 20.4 Steps in plain concrete strip foundation.

need to be taken into account beyond around 1200 mm below ground level.

To limit the depth on a sloping site the foundations can be stepped as shown in Figure 20.4, where the dimensional requirements are for a plain concrete strip.

Foundations for poor surface conditions

Apart from sloping sites, other conditions lead to the need to consider deeper founding depths; these include:

- Potential volume change in shrinkable clays
- Variability of the soil across the site
- Surface soils with poor bearing capacity

The production costs and risks will have a great influence on devising solutions to provide deep foundations. Indeed the answer may not be to provide deep foundations at all but to support the building at the surface in such a way as to limit the pressure on the ground and stiffen the foundations to limit the effect of differential settlement on the structure and fabric of the building.

Solutions available fall into three categories:

(1) Foundations to depths around 4500 mm
(2) Foundations to greater depths
(3) Surface 'floated' rafts

Deep foundations

Solutions for depths to 4500 mm have been mainly developed for sites with problems

associated with shrinkable clay, but they can be considered for any sites with good bearing ground within this distance from the surface.

In clay soils, or more generally soils that can be excavated without the need for earthwork support within the first few hours of excavation, trench fill foundations can be considered. These are shown in Figure 20.5 where, however deep the founding level, the concrete is filled up to a level that allows the safe and economic laying to the wall up to the DPC. In this type of foundation the production conditions are the key. The trench has to be able to be excavated accurately (line, level plumb and width) without the risk of collapse. This has to be achieved in a reasonable time to excavate sufficient length of foundation, confirm the founding depth and pour the concrete to return the ground to a safe condition.

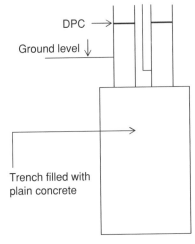

DPC

Ground level

Trench filled with plain concrete

Figure 20.5 Plain concrete 'trench fill' strip foundation.

Generally this cycle of operations will have to be completed within one day. These conditions can be fulfilled in most clay soils, but the time factor may limit the depth to which this form of foundation can be undertaken, as the longer the trench is left open the greater the risk of collapse as the moisture content of the clay changes. As long as the trench can be cut cleanly and accurately the quantity of concrete can be controlled, limiting the financial as well as the health and safety risks.

If there is then a risk of heave (expansion) in the mass of soil contained under the building between the foundations, then a compressible material should be introduced to the face of the foundation, and the ground floor possibly constructed with a void as a suspended floor.

An alternative to trench fill is to provide piles to take the loads down to the safe bearing soil with a beam across the top of the piles to support the wall. Up to depths of around 4500 mm this can be achieved with what are known as short bored piles, as illustrated in Figure 20.6. Here short and bored are determined by the production requirements. Piling is achieved either by removing the soil and then replacing it with concrete (replacement piling) or by driving the pile to displace the soil (displacement piling), and these can be achieved in timber, steel or concrete. In replacement piling the method of removing the soil is by boring or drilling a shaft into the ground. This process will produce a circular shaft, the diameter and length of which will determine the size of plant (piling rigs) that will be required. By limiting the length to around 4500 mm and the diameter to around 350 mm, piles can be achieved with relatively small rigs. If greater depths and/or larger diameters are required then full piling rigs are required.

Even with a circular shaft with a limited diameter, some soils may collapse into the hole before there is a chance to fill the shaft with concrete. In this case the shaft must be created with a tubular lining normally recovered as the pile is concreted. Displacement pile systems with precast piles and beams have also been developed where this risk of collapse is considerably reduced so long as founding depth is well established and verticality during driving can be achieved. Both of these conditions can usually be found on clay sites.

The beam has to span between piles to support the wall. Piles are placed at relatively close centres of around 3 or 4 metres, limiting the need for a deep beam. In addition, piles are placed at corners of the building and at the ends of door and wide window openings. Production operations to create the beam can be achieved

DPC

Ground level

Depth to stable ground of adequate bearing

Reinforced concrete ground beam

Bored, cast in-situ piles

Note precast pile and beam systems can be used

Figure 20.6   Pile and beam foundation.

in a trench near to ground level. Here the beam has to be wide enough to carry the wall so the characteristic shape of the beam is wider than it is deep, not the usual proportions of a beam. If these foundations are in shrinkable clay with the beam near the surface, then shrinkage and heave could take place under the beam and around the top of the pile. The beam must be cast on a compressible layer and in some cases there may need to be a slip membrane around the top section of the pile.

Where founding depth is greater than 4500 mm then the need for longer piles necessitates the use of full-size piling rigs, but the principles are the same. With the relatively light loads from houses, the need for deep piling is unusual. There are a number of options of both replacement and displacement systems to create the piles. A full discussion covering the choice of deep piling is beyond the scope of this case study.

The idea of the short bored pile can also be achieved with a pier and beam foundation. This involves the excavation of pits for pad foundations where masonry piers are constructed to take the beams that support the wall. This requires men to work at the founding level with the necessity to provide working space and health and safety measures.

Raft foundations

The alternative to providing deep foundations is to investigate the use of surface 'floating' rafts. These combine the ground floor and the foundations into one continuous structural unit. All the loads combine and are spread out under the raft so the entire footprint of the house is the foundation bearing area. This will create very low pressures on the soil under the raft, although it will stress the soil to a greater depth, albeit that they will be low stress values. This gives a safe solution on soils with limited bearing capacity and has the potential to accommodate differential settlement without transferring the strains into the superstructure. If differential settlement occurs, the raft can be designed to tilt as a whole with no strains induced in the superstructure (this can be successfully used in areas subject to

mining subsidence). This can only be achieved if the pattern of loading is relatively even to give the centre of gravity of the loads at the centre of the area of the raft and the construction can produce a raft stiff enough to act as one foundation.

The stiffness or rigidity of the raft is the key on sites where there is variability in the soil strength and/or compressibility under the building, giving the possibility of differential settlement. As the foundation to a house is a limited area with a relatively high perimeter to floor area ratio, it should be possible to achieve this rigidity by stiffening the edge where the main loads will be taken from the external walls. Thickening may be necessary under internal loadbearing walls. As the ratio of perimeter to floor area decreases, it may become economic to stiffen the internal area of the slab with beams rather than just increase the thickness of the slab.

For housing, the slab is likely to be between 150 and 300 mm thick with an edge stiffening detail dependent on the soil conditions. For soil of medium to low compressibility that is consistent under the whole raft, the edge detail may only need to be a reinforced extension beyond the external face of the wall. These details are shown in Figure 20.7. This extension is particularly significant when the raft is constructed close to the surface. Relatively small extensions can give some protection from frost heave but may have to be greater if the raft is on clay, with the risk of shrinkage and loss of support under the edge of the raft. On granular soils a small downstand provides the best protection, as erosion may progressively undermine the edge of the raft over time.

On soils of high compressibility but particularly where the soil is variable over the site, effective support may be reduced over some sections of the foundations, and there is a need to provide additional bending capacity as well as protection. Here the downstand beam detail shown in Figure 20.8 will be required.

All rafts are reinforced with mats for the slab that is usually provided with preformed mesh reinforcement. For light edge downstand details, pre-bent mesh cages may be considered, but for edge beam raft design the beam cage

Figure 20.7 Plane raft foundation.

Figure 20.8 Edge beam raft foundation.

will need to be specified with bars and links that may be fabricated on site. As with the 'wide' strip foundation, the reinforcement will need to have sufficient cover, and to ensure this is maintained in the production process, there is a need for 50 mm of concrete blinding and the use of spacers and chairs to maintain the position of the top mat in the slab.

Economics of foundations

The evaluation of costs for foundations with the elements of risk and variety of site conditions can be a major determinant of the choice of foundation method. Depth is not the only factor. If the production process for deep foundations can be carried out from the surface with little risk, they may be less expensive than a raft with formwork, reinforcement and concrete making up the major cost. For the strip foundation options, when to use wide strip or trench fill rather than the simple strip will again be dependent on site conditions and the risks involved.

To this might be added the possibility of using ground improvement techniques to increase the bearing capacity of the soil particularly in non-cohesive soils. This may make simpler, less expensive foundation viable, offsetting the cost of the ground improvement process. Techniques such as dynamic consolidation and vibrocompaction and vibroreplacement can improve ultimate bearing capacity to economic depths for strip foundations to be used for housing.

Basements

Basements have not found favour with UK households, where underground rooms are more associated with cellars and coal holes than habitable rooms. The construction of the basement has to deal with horizontal ground forces as well as the loads from the construction above. It also has to have high performance levels of waterproofing, which have been poorly achieved in the past. There are now established construction methods that can achieve high levels of performance, but they are beyond the scope of this case study.

Summary

(1) The basic function of the foundations is to maintain the integrity of the structure and fabric of the construction above ground.
(2) The foundation has to take account of the pattern and size of the loads from the superstructure and the stiffness of construction above ground as this determines its ability to respond to differential settlement.
(3) The soil needs to be investigated to determine not only the structural properties of the soil but also its moisture stability and any contaminations over an area and to a depth governed by the scale of construction and any suspected natural and man-made features that might affect the stability of the foundations.
(4) Foundations will have to have sufficient area, have loads placed centrally, be stiff enough to transfer loads evenly to the ground and be at sufficient depth to limit settlement and avoid collapse.
(5) The simple strip of plain unreinforced concrete will be sufficient for most houses considered in the case study.
(6) In site conditions where the construction of the simple strip carries higher risks and costs due to the required site operations, trench fill, piles or a raft 'floated' on the surface can be considered.

21 Services

This chapter looks at the active technologies, the building services that provide environmental and operational support to the household. It provides the analysis for a number of domestic systems, with particular reference to the range of emerging systems to support the growing environmental concerns of sustainable development.

Introduction

Chapter 10 discussed the requirements for the choice of construction to create and maintain environments in buildings. It introduced a group of technologies that are active in this role which are part of a range of systems known generally as building services. These technologies have grown in importance in the contribution they have made to achieve environmental performance levels over the last 100 years. They are now subject to further changes from the growing need to consider sustainable development as a factor in making choices. They significantly affect the use of energy and other resources such as water, as well as producing waste and potential pollution during the occupation of the building. These are typical factors being used to measure the environmental rating of a building.

This evaluation against sustainability criteria is leading to changes at all levels of analysis of services. It has involved changes in the efficiency of part of the system, as with central heating boilers, as well as the introduction of completely new systems, as with solar heating and grey water systems, which will be discussed later in this chapter. Perhaps the most radical has been the change in design of the whole building, integrating the active services with both the passive and active fabric, as with the design of natural ventilation systems explained in Chapter 10. With the use of motorised windows and blinds in office developments the distinction between active services and fabric is becoming blurred.

For the house being considered in the case study the most likely services that would be included are identified in Table 21.1. All these systems contribute either directly or indirectly to the quality of the internal environment to maintain the well-being of the occupants as well as the objects and equipment in the house, as discussed in Chapter 10. However, in this text only four systems will be examined in detail:

- Hot and cold water
- Heating
- Drainage (above and below ground)
- Lighting and power

Analysis of services systems

It was suggested in Chapter 10 that services have to be thought through in a different way

Table 21.1 Typical domestic services systems.

Supply (utility) services	Input services systems	Disposal services systems
Water	Heating	Drainage
Electricity	Hot and cold water	Air extract
Gas	Lighting	
Telecommunications	Power	
	Security	

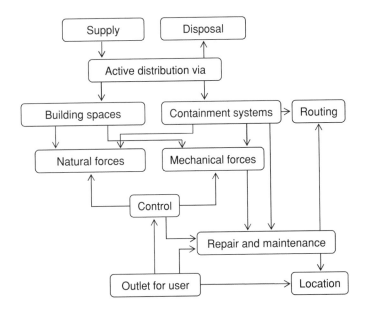

Figure 21.1   Aspects of active services systems.

to the passive fabric. The aspects identified are shown in Figure 21.1, a repeat of the figure given in Chapter 10. The essence of a building service system is active movement. In the environmental services this involves either supply or disposal in order to maintain a safe, healthy and comfortable environment for the user. These are shown at the top of the diagram. The system must provide active distribution via a containment system (pipes, wires or ducts) or the spaces in the building itself using either natural or mechanical forces. Once a basic system is established then control has to be chosen and decisions taken on the provision to effect repairs and maintenance. It is then possible to think about location of the parts and establish the routing of any containment systems within the building.

As indicated at the beginning of this chapter, beyond the aspects shown in Figure 21.1, cost and sustainability are significant at all levels of choice, influencing both details and the introduction of completely new systems.

Hot and cold water

From the assessment of the source, through the whole process of choice associated with water supply and distribution systems, two factors dominate the details and specification:

- Quality of water
- Quantity (rate) available

The choices are made within the context of resources, cost and environmental demands, but the impact of these has to be assessed against the criteria of quality and quantity to ensure that basic health and hygiene standards are maintained. Without water of the right quality in sufficient quantity, sustainable development cannot be achieved. This may involve education and a reassessment by users of their expectations and their behaviour that sets performance level, but the supply of clean water remains fundamental to maintaining healthy communities.

For each household it is possible to identify at least four different quality categories, only some of which have to be provided as hot as well as cold water:

- Drinking and food preparation
- Cleaning and personal hygiene
- Disposal of waste
- Use in services systems

Drinking and food preparation (potable water) is likely to be the highest standard and therefore the source must be of this quality. For most households in this case study the source will be from a mains supply through a pipe laid in the street outside the property, but there are still some wells and springs used in rural areas in the UK. All sources are subject to regulation to maintain their quality. Typical measures of quality will include standards related to:

- Biological content
- Chemical content
- Particulates
- Appearance
- Taste

Consideration of this list will show that while all of these need to be of a high standard for drinking water, lesser standards may be possible for the other uses of cleaning, disposal and services systems. Some care is needed in setting these 'lower' standards, as exemplified by the incidents of legionnaires' disease. Spread by droplets of water in the air, two sources have been identified. From air conditioning cooling systems it can be controlled by disinfectant, as the water does not have to be drunk or otherwise come into contact with the skin. However, the same pathogen spread in the droplets of water from showerheads has to be controlled by heating the water as the quality is associated with personal hygiene rather than a service system.

Even the regulation associated with the quality of the source may not provide the quality required by the user. Many households will filter the water for drinking (or buy bottled water) and will have to reduce the calcium carbonate in hard water areas for any water that will pass through an instant hot water heater in order to limit the build-up of scale on the heat exchanger. Both of these are primarily to change the chemical content of the water.

Water for drinking and food preparation for a house has to be supplied to the kitchen. This must be taken directly from the supply via a distribution system that will not contaminate the water or allow cross contamination from other parts of the system carrying water of a lesser quality. Many of the regulations governing the materials and specification of these systems are concerned with contamination and cross contamination. There are many requirements and it is not the intention of this text to give them all, but an example is the requirement of the use of leadless solders for any jointing of the pipe work that is to carry water of this quality. Lead is one of the chemicals controlled at source as it is harmful to health; any use of lead (which is soluble, particularly in soft waters) in the system may compromise this standard.

It is common practice (although by no means universal) to store water in a tank (cistern) normally in the roof space. This water cannot now be considered to be of drinking quality but may be used for all the other water needs of the household. Still some control is necessary, such as covering the tank and maintaining a minimum distance from the inlet to the water level to stop cross contamination.

Storing water in a tank has another effect, not on the quality but on the potential quantity available at the tap or other outlets. Important to the user is the rate of delivery, or how long it takes for the quantity required to be delivered at the outlet. The quantity passing through the pipe is measured in litres per second. This will be dependent on the size of the pipe and the pressure that is providing the force to move the water through the pipe work distribution system. From the mains the water authority has to provide the water at a pressure within limits to ensure an adequate rate of delivery. Once this water is stored in a tank the pressure comes from the level of the tank relative to the outlet. This is why the tank is in the roof space to give a reasonable pressure to outlets on the upper floor. Generally the pressure from the tank will be lower than that from the mains. Mains water is said to be the high-pressure system and water from the tank on the low-pressure system. This is why washing machines and dishwashers may have to be taken off the rising main. They do not need the quality of water but need the pressure to operate successfully.

Basic system

The system identified so far is shown in Figure 21.2. It has two pressure systems and only two qualities of water. Hot water is provided from a tank, although the heating source is not yet identified. Hot water is only required for cleaning and personal hygiene so can be supplied from the tank on the low-pressure system.

Figure 21.2   Hot and cold water – basic system and pipe work.

The figure shows the basic parts of the system:

- The supply identified as an incoming main
- Range of outlets required: sinks, baths, toilets, etc.
- The pipe work and tanks of the distribution network
- Operational control arrangements: taps and float valves

Analysis of the rate of delivery at the outlets shows that the greatest rate is required to the WC pan where the flushing effect requires a relatively high volume in a short space of time. Even on the high-pressure system this would require a large pipe to be able to achieve the cleansing and disposal effect required using a wet WC system. The alternative that is shown in Figure 21.2 is to deliver the water to a cistern at a relatively low rate to be stored and released in a short space of time to flush the pan.

The outlet requiring the next highest rate is the bath. Due to the large quantity required to fill a bath a reasonable rate of delivery is required. This can be achieved, even on the low-pressure side, with a 22 mm pipe. All other outlets can be supplied using 15 mm pipes including the WC cistern. Pipe work of this size can be created using copper pipes that provide a safe and economic solution, having a wide range of fittings to configure the network.

Within a house this system should be able to maintain a rate of delivery, even if more than one outlet is being used, but performance will drop as more outlets are in use at the same time. All systems have to be designed for some simultaneous demand, but it is unnecessary to design for all outlets to be used at once, as the probability of this happening is very low. However, different systems will have different capacities to provide for simultaneous demand and this must be one of the criteria against which performance will be judged.

The tap provides basic manual control for hot and cold water systems, although cisterns need to have a float (ball) valve to cut off the water when the required level of water in the cistern is reached. Electrical appliances such as dishwashers and washing machines will have electromechanical devices to admit water at appropriate points in the cleaning cycle.

Maintenance, location and routing

Once the basic understanding of the system, its supply, outlets, active force, distribution system and operational control, has been identified, it is necessary to consider the following:

- How repairs and maintenance will be accommodated
- The location of the outlets and the routing of the distribution

The system considered so far, even with its control for everyday operations, will be difficult to maintain and repair once the system is full of water. Essential to any work that has to be undertaken on the system once it is in operation is the need to isolate sections of the system by turning off the water in order to carry out the works. It is then an advantage to drain off the water in a controlled manner so that the pipes are empty before the work starts. How small a section of the system needs this individual isolation is the next decision. When systems were first introduced into houses it was common practice to provide only two points of isolation and often no facility for draining down the pipes in a controlled manner. These would have been a stop valve in the pavement outside the premises, belonging to the water supply company, and a stop valve on the incoming (rising) main as it entered the house, belonging to the owner of the house. If a storage tank was provided, a gate valve would normally be installed to isolate the low-pressure side. These two stop valves (and the gate valve from the tank) are still the primary points of isolation for a hot and cold water system. The water authority's valve is also now used for any metering and the valve on the rising main will have a drain tap for controlled emptying of the system.

It is now good practice to provide more isolation points. Deciding on how many isolation points and where they should be comes from an analysis of where there will be a need for maintenance (and how often) and the need for fast

isolation in case of failure, to minimise damage before a repair can be arranged. There is then a consideration of keeping as much as possible of the system operational while the maintenance or repair work is carried out on defective parts.

A typical arrangement of isolation and drain-off is shown in Figure 21.3 superimposed on the basic system. The most common maintenance requirement is to change the tap (or cistern float valve) washer and therefore an isolation valve is installed just before the tap or cistern. This allows the work to be achieved quickly without affecting any other part of the system or having to drain down any water. Having isolated each tap and cistern, more major section isolation could be considered. If a tank is provided in the roof space it will include the gate valves from the tank to the low-pressure systems for hot and cold water. It may then be worth considering isolation and drain down for each room (kitchen and bathroom) for more major works in these areas, such as future refitting.

These points of isolation must be given good access, as must the items that need maintenance and/or replacement. This is particularly true of any isolation that has to be reached quickly in the case of failure where water damage needs to be minimised. This has to be taken into account in the location and routing decisions for the system.

In contrast to this need for access is the requirement for concealment that comes from a concern for appearance. It is now common for householders to wish not to see the services, so concealment becomes another key aspect of the location and routing decisions.

Concealment cannot be separated from space competition, as the most obvious space for concealment is the construction zones where structure and fabric for other performance requirements are already taking up space. Any distribution that cannot be accommodated in existing construction zones will have to take up useable space in the building, altering the room size and shape.

The two large items in the system are the hot and cold water tanks. The location of the cold water storage cistern has already been identified

Figure 21.3 Hot and cold water – isolation and drain down.

as being in the roof space in order to provide a sufficient head to create the pressure to maintain a required rate of delivery to the outlets. The gate valves for the low-pressure hot and cold water systems from the tank will therefore also be in the roof space. The location of the hot water tank, however, can be at any level as it is the location of the cold water storage cistern that will dictate the pressure (less pipe losses, but these will not be significant in a small system for a single dwelling). The hot water tank needs to be fairly central, often in a cupboard somewhere in the house, to ensure no outlets are at the end of long pipe runs, thus avoiding excessive delays in the hot water reaching the tap. If a roof space is not available for the cold water cistern and space is limited to provide a central cupboard for the hot water tank (common in flats and increasingly the case in houses), systems can be specified that eliminate the need for one or both. One such system is the use of a combination boiler, which will be introduced later in this chapter with the heating system.

The pipe work system, while taking less space than the tanks, needs to be more widely distributed. It is now necessary to take account of design and operational efficiency. The grouping of outlets to minimise pipe lengths and bends reduces losses in the system and the cost of pipes and fittings. This is more of a consideration in a large commercial installation. In housing, the drainage system will have a larger influence on the need for grouping or stacking wet serviced areas (kitchens and bathrooms), but this will tend to also concentrate the hot and cold water supply systems.

Even with this concentration of wet areas, some vertical and horizontal distribution will be necessary. Options for horizontal distribution have already been considered in Chapter 17 where the use of the floor void was discussed. There are limits to where joists can be cut for pipes to pass and the introduction of sound insulation has implications for the free space available in detailing to ensure that sound performance is maintained. Vertical distribution may have to be boxed into ducts in the corner of rooms. They may have to be buried or hidden behind finishes, but this will make access more disruptive and accidental damage more likely.

Life cycle and sustainability

Because of the active nature of services, life cycle issues like cost and sustainability are particularly important considerations. There is a need to consider the life cycle in the following stages:

• Design
• Installation (production)
• Operation (running)
• Replacement (maintenance, repair and refitting)
• Disposal

For cold water, the use of a natural pressure head means that running costs in fuel and energy terms are low, but there is a need for saving water, particularly high-quality drinking water, as an expensive and precious resource. Saving water is worthy of consideration even if it increases installation requirements. For hot water, the heating options do have fuel and energy implications.

There is a concern for the increasing demand for water and the impact of its disposal after use. Both are incurring increasing infrastructure costs that have to be passed on to the consumer and are having a high impact on the environment. The choice of system for each individual house can make a considerable difference to future costs and impacts.

Designing systems that use less to do the same job will contribute to sustainability and this can be exemplified by the provision of showers (to limit the need for baths) and in the use of low volume, dual flushing cisterns for WC toilets. While this reduces the quantity of water used they may still use water of a high quality even though this is not strictly necessary.

It has been recognised that the volume of water used for cleaning and personal hygiene by the average household equates with the volume of water used for flushing the WC(s). Grey water systems intercept the cleaning water before it enters the drains. The used water is filtered for particles and other suspended matter that affect clarity plus some disinfecting and then pumped to a separate tank in the roof. This is then connected to the toilet cisterns to be used for flushing. This has to be a parallel system with the ordinary water supply, as, for health reasons, the household cannot be left without flushing water if the grey supply cannot be maintained. Similar systems can be installed to capture (harvest) rainwater which needs less cleaning, but the system has to accommodate much larger storage provision in order to smooth out supply (rain) and demand (flushing WCs). These grey water systems clearly increase installation and maintenance costs, but, with metered water supply, will show a saving in running costs and will have a low impact on the environmental needs for both water sources and disposal.

For the running of hot water, solar heating, even at the latitude of the UK, can make a significant contribution to the annual requirements for heating water. This would have to be a dual system with conventional heating sources, but again it can save on both running costs and environmental impact.

These examples are of system choice that affects the operational use of water and energy and can therefore make a major contribution to sustainable development. Using less, recycling and using renewable energy sources achieve this. All these have the effect of reducing pollution and helping to maintain habitat by reducing overall demand on the infrastructure. Some of these could only be achieved by introducing new systems that have to be accommodated in the building, paid for at the time of construction, maintained and disposed of at the end of the life of the system. The initial cost can often be recouped over the life of the system if water is metered, but the environmental impact, significant to the community and future supply (and hence cost), is less easy to calculate for each individual household. The choice comes for a commitment to all aspects of sustainability, economic, social and environmental, and not just the notion of an economic life cycle cost.

There are sustainability issues in the choice of materials and components for the actual systems, their manufacture, transport and disposal demands. The sanitary fittings will be stoneware and the plumbing systems are dominated by metal components, mainly copper, although plastic can be substituted for some of the components. Analysis of these is important, but currently systems that save water and fuel and have lower demands on the sewer system are seen as making the largest contribution to sustainable development.

Heating systems

There are a number of ways of providing heating for a house. This can be based on individual heaters in some or all rooms, perhaps with a variety of heat sources and/or emitter types to match the occupation pattern. However, in the case study being considered in this text, a full central heating system is likely to be the expectation of the user, with high levels of insulation to reduce component sizes and running costs. The most usual expectation of a central heating system is radiators as heat emitters in each room, supplied with hot water pumped through pipes heated in a boiler most probably fired by gas, if this is available.

In a domestic central heating system the main control will be based on a time clock (to set heating periods) and a single room thermostat to turn off the boiler (and pump) when the house is up to temperature during the required heating periods. Some of the individual radiators can then be fitted with thermostatic valves to give control over temperature in some rooms during heating periods in order to match occupation pattern. Even where thermostatic valves are not fitted, there will be a valve on the flow to each radiator for the occupier to turn the radiator off completely by shutting off the flow of water. There will also always be a valve on the return side of the radiator, but in this case it is not a control for the day-to-day operation of the system. Known as a lock shield valve, its purpose is to balance the system to ensure a fair distribution of water to all the radiators and is set at the time of

commissioning when the system is first brought into operation. The need for control, not just for operational purposes but also for commissioning, is a feature of many services systems.

This expectation for a central heating system of the form outlined above is the case for the mass housing market, although alternatives, often based on more sustainable solutions, are emerging for individually designed houses.

The description of the typical central heating system indicates five major parts to the system:

(1) A heat source – the gas-fired boiler
(2) Heat emitters – the radiators
(3) A distribution network of pipes carrying water
(4) A pump to move the water (heat) from boiler to radiators
(5) Control from time clocks, thermostat and radiator valves

Figure 21.4 illustrates a basic system fed with water from a low-pressure feed and expansion (of the water when heated) tank in the roof space. This also shows that the same boiler will heat the hot water in a hot water storage tank. However, because the pattern of demand for hot water is different to heating it will require separate demand (motorised valve) and temperature controls (tank thermostat) and may even have an alternative heat source such as an electric immersion heater. This also complicates the control of the boiler which is normally resolved in the programmer. The pipes are shown as a two-pipe system with a flow (to the radiators from the boiler when the water is hot) and a return (back to the boiler after the water has cooled, having been through the radiator). The pump will maintain an adequate flow around the system. On a domestic system such as that shown in Figure 21.4, the pipes will be 15 mm diameter up to each radiator, with 22 mm pipes for the flow and return, carrying the water from several radiators to and from the boiler.

Maintenance, location and routing

In the system shown in Figure 21.4 the most likely parts to require maintenance and replacement are the boiler and the pump. Not only does

Figure 21.4 Central heating – conventional system.

this mean that access must be available, but, in the case of the pump, isolation valves on either side allow it to be removed without draining down the system. The radiators may need to be removed, not for wear in the system, but for changes and redecoration of the room. The valves that are provided at either side of the radiator for control can also be used to isolate the radiator so it may be removed without draining down the whole system.

The location of the boiler has to take into account a number of factors. There is a need to connect the boiler to both a fuel source and a flue to safely take away the waste products of combustion. Some types of flue can be restrictive. Boilers with balanced flues have to be on an outside wall, but fan-assisted flues can take the gases horizontal for a limited distance, allowing a more internal location for the boiler. The concentration of pipes and connections around the boiler requires space and may be best concealed in some rooms but may be left exposed in others. Boilers do create noise directly from the unit and the noise may be transmitted via the pipe system. This should also be taken into account in locating the boiler.

The other parts of the system requiring careful consideration for position are the radiators. Their position may have some effect on the comfort of the individuals using the room. Radiators act mainly as convectors but do have a radiant component of heat exchange. The convector action promotes a circulation of air in the room to warm the space. The radiant heat acts over a limited distance, as the surface temperature of the radiator is limited for safety reasons. However, its influence, both radiant and the warm convective current rising above the radiator, is opposite to the cool feel of the windows with falling cold air when temperature is low outside. Although this is less important with double and triple glazing, the positioning of the radiator below windows tends to balance the effect of the cool glazing, so this has been the traditional position for radiators. It is also true that there is a convenient wall space below the window that is less useful than other walls in the room. With higher levels of insulation and smaller radiators, this position is less critical so long as a convective current can be achieved and no sedentary activity is too close to the radiator where the radiant component can be uncomfortable.

With radiators in each room, particularly if located under each window, the pipes will be widely distributed. Vertical distribution is likely to be above the boiler and horizontal distribution in the floor, but on a pumped system this is not critical. There will probably be a preference to conceal pipes, but some discreet exposed pipe work may be acceptable. All controls and valves must be easily accessible.

Life cycle and sustainability

The five major phases in the life cycle (design, installation, operation, replacement and disposal) were identified in the section on hot and cold water above. In contrast to the water systems, the heating system is a major consumer of energy. This not only affects the economics of the choice of system over the life cycle but also has major implications environmentally.

Consideration of life cycle issues has to take into account the specification for the passive construction as well as the specification of the active heating system. Both insulation and the heating system itself have a capital cost and affect running costs and energy consumption. Concern is now not just for the cost of the fuel but for the type of fuel and its environmental impact, particularly carbon dioxide emissions as a major greenhouse gas. Electricity produced from renewable, non-polluting, safe sources available in some countries in the world would represent low carbon dioxide emissions. However, this is not the case in the UK at the beginning of the twenty-first century where this case study is set. With limited options on energy sources that are not based on burning fossil fuels, the emphasis is on minimising the loss of energy in the conversion process from burning the fuel to delivered heat. It is worth noting that even the non-polluting, renewable energies may be based on large infrastructure projects, so minimising the use of energy for any individual building should still be a key objective.

Insulation is recognised as the best way to save energy. Well-insulated designs can, by taking into account heat inputs from solar gains and the occupants and equipment in the building, pro-

vide homes that need heating on fewer days per year and/or less hours in the day. However, heating systems' efficiencies can also help in the move towards lower energy consumption when the heating system is operating. The use of thermostatic valves in some rooms that are less used or do not need such high air temperatures is an example of a system control measure that saves energy. Boiler efficiency is another way of minimising energy consumption. Boilers are now rated for energy efficiency, where the best gas boilers are currently based on condensation mechanisms where heat is recovered from flue gases to increase the effective heating obtainable from each cubic metre of gas. Running systems at higher flow temperatures is more efficient to increase transfer of heat into the room by both convective and radiant processes, but this makes temperature of the surface of the radiator high, introducing safety and comfort considerations for anybody who gets close to the radiator.

Given users' expectations for central heating in their homes, the dominance of the gas-fired system is due to the current economic balance between capital and running costs. While more expensive to install than the systems based on electricity, the running costs are lower with the current cost of gas.

Alternative central heating installations options

Installation costs (and possible running costs associated with water heating) can be reduced with the use of a combination boiler. The central heating is now on the high-pressure side of the water supply and the boiler becomes an instant hot water heater. An installation using a combination boiler is shown in Figure 21.5. This system does not need a low-pressure feed and expansion or a hot water storage tank. This also simplifies the control (programmer) as the hot water is now on demand from the taps and only the heating has to be controlled from the clock and room thermostat, independent of the hot water.

Combination boilers were developed for flats where a feed and expansion tank was difficult

Figure 21.5 Central heating – combination boiler.

to position, occupation was often for few people and distances to hot water taps were short. While hot water is always available (no time delay in reheating water in a tank once it has been used) it is not so good at simultaneous demand, as the boiler can only provide a limited quantity (litres per second) at the required temperature. If the boiler is not central then it may take some time for hot water to reach the tap, wasting both water and energy to heat the water left in the pipe to cool.

To overcome these problems with the hot water in larger households, high-pressure boiler systems (no feed and expansion tank) can be used to heat a tank of hot water. These are known as system boilers and need to be provided with the control and programmer facility of the original low-pressure installations. The hot water is often still a low-pressure system being fed from a cold-water storage tank in the roof space. However, it is possible to use unvented cylinders connected to the rising main, making the whole installation high pressure and taking away the need for a cold-water storage tank in the roof space for the hot water supply.

Drainage

As a disposal system, drainage is concerned specifically with removing unwanted water from the house. Drainage deals with unwanted water from two sources:

- Foul water: comes from cleaning and personal hygiene that would pose a direct health hazard and a threat to the fabric of the building if it were not removed
- Surface or rainwater: collects on hard surfaces such as roofs and paths which if not removed increases the possibility of damp conditions that would threaten the fabric and create potentially unhealthy environments

Both of these have systems above and below ground that, while based on similar design principles, will require different materials, jointing and support arrangements for the pipes with different maintenance and routing considerations.

This text will introduce the principles and consider foul water drainage systems. Surface water or rainwater systems are almost identical, although the risks of escaping odours and blockage are less.

'S' trap   'P' trap   Bottle trap

Figure 21.6 Trap fittings for plastic waste systems.

Basic system

As the drainage system takes water from collection points above ground to pipe work below ground it is possible to use gravity to move the water away from the building. In nearly all cases the final discharge point is low enough compared with the levels around the house to arrange for the flow of water from collection to discharge without any resort to pumps or ejectors. To achieve this flow, pipes have to be laid to a fall or gradient to keep the water moving through the pipe.

Pipes do not run full but act as channels carrying the water and any solid material that is present in the foul water. The air that remains in the pipe will become unpleasant, with a smell, generated by the waste, which will accumulate as a gas in the pipe. The system has to ensure that the solids are carried away and that the gas is not released where the smell would be offensive.

To carry the solids, the water must maintain a self-cleansing velocity. If this velocity is not maintained the solids are not washed away by the water and remain in the pipe, building up and causing a blockage. Typically, for a domestic underground drainage system from a single dwelling, flow rates are very low, so the minimum pipe size for foul water of 100 mm would be specified and a fall of around 1:40 (1 metre dropped for every 40 metres travelled) would be required. Shallower falls would be possible but rely on greater accuracies in production to ensure falls are maintained throughout the pipeline. As well as pipes with these relatively shallow falls, a vertical pipe is also self-cleansing. Drainage systems are, therefore, made up of pipes with relatively shallow falls and vertical drops to connect sinks, toilets and other appliances to the disposal system.

Gas (smell) will be released back into the room if the connections that are made at the appliances are made straight into the drain. This escape of gas back into the house can be stopped with the simple device of the water seal. The seal is moulded into appliances such as toilet pans but will have to be provided by fittings known as traps under sinks and other waste outlets such as washing machines and dishwashers. Trap fittings for plastic waste systems are shown in Figure 21.6. The bottom section of these traps can be removed for cleaning, as traps are prone to blockage.

The act of sealing the system at the point of collection creates the conditions for differential pressures to be generated in the system when water is flowing away. These conditions tend to remove the seal when the system is in operation. This can happen to both the seal at the appliance being used (self-siphonage) and then to other seals as the water passes through the system (induced-siphonage and back pressure).

It is necessary to have the system open or vented to normal atmospheric pressure so that these operational pressure changes are minimised and quickly restore an equal pressure on both sides of the water seal. The opening to the air also lets the gas out of the pipe work and has therefore to be located where the smell cannot be offensive to anybody either using or passing by the house.

Soil vent pipe (SVP) →

Example of design limits for single stack to bathroom appliances				
Appliance	Pipe size	Max. run to stack	Gradient	Seal depth
WC	100 mm	6 m	9 mm/m	50 mm
Wash basin	32 mm 40 mm	1.7 m 3 m	Varies 120 to 20 mm/m*	75 mm
Bath	40 mm 50 mm	3 m 4 m	Varies 90 to 18 mm/m*	75 mm
* Larger gradients for shorter lengths, e.g. basin 50 mm/m for 1 m run				

Long radius bend to underground drainage

Figure 21.7   Above ground drainage – domestic single stack system.

Above ground system

A typical above ground domestic installation known as a single stack system with pipes, seals and venting is shown in Figure 21.7. In this system each appliance is individually connected to the main vertical stack. There are limits on distances and falls related to the diameter of the pipe and depth of seal in traps, which minimise the risk of siphonage, examples of which are given in the table in Figure 21.7. Making the main stack a soil vent pipe is sufficient to prevent siphonage, with pressures being restored quickly enough, so long as the basins and baths have relatively flat bottoms for the last of the water to drain slowly to refill any loss of water in the seal due to siphonage. In these domestic systems, venting of individual appliances is not necessary.

The pipes will most probably be uPVC, although cast iron may be used for the vertical stack if there is any danger of damage, perhaps if the pipe is located in a garage. Plastic is easily cut and will have a range of fittings such as traps, bends and connectors to the stack, with joints that can be formed either with cold solvent welds

or, more usually, with rubber ring seals with push fits or hand-tightened collar nuts. The traps in Figure 21.6 are illustrated with hand-tightened collar nuts.

Below ground systems layouts

While above ground single stack systems will be similar in nearly all houses, underground drainage is governed by the site and the availability of a discharge point. The pipes still have to have limited falls or be vertical to be self-cleansing, but it also becomes important to think about ensuring a stream line flow at bends and connections so as not to stop or slow down an existing flow at bends or when a second discharge joins another pipe. With limits to falls vertically and consideration of bends and junctions on plan, this makes the pipe system a three-dimensional layout below ground. It is important to relate this to ground level to ensure an economic system, as depth below ground has a considerable influence on the initial cost and on the ease and cost of any subsequent maintenance and repair.

It is necessary to identify how and where the waste can be safely discharged into an existing system. For most development this will be a public sewer located in the road outside the property. If there is no sewer available an economic distance from the building then cesspools (holding tanks below ground on the property that need to be emptied when full) or in more rural areas septic tanks (where some treatment of the liquid takes place) can be considered. Small-scale treatment plants can also be considered for isolated single or small group developments where road access to clear cesspools and septic tanks might be difficult.

Even if public sewers are available, the systems may vary. The most common is a separate system where foul and surface water have to be discharged into separate sewers, creating the necessity for two separate underground drainage systems to be installed on the property. Older sewer systems may be combined, taking both foul and surface water, or even partially separate, where independent foul and surface water systems have to be created on the property but a limited amount of surface water can be put into the foul drain to simplify layouts.

An example of a below ground foul system for a single house is illustrated in Figure 21.8. The plan shows two above ground stacks collecting from bathroom and kitchen appliances and a utility room with sink and washing machine connections. The illustration assumes a public sewer on a separate system, although the surface water drain is not shown.

The need to maintain a streamlined flow at bend and connections to minimise the risk of blockage at these points is achieved by taking all the pipes and connections towards the sewer. No connections or bends go against the flow. Indeed, all the connections other than the short connection from the utility room has pipe bends and connections at angles greater than 90°, the angle being influenced by the fittings available.

The frequency and position of access is part of the decision making for the pipe layout. While the streamlining of flow is to minimise the risk of blockage there is still a risk often associated with misuse (flushing items into the drains for which the system was not designed). While even this may be a rare occurrence, the consequences are unpleasant and place a high risk on health. Provision has to be made for the speedy clearance of a blockage with minimum disturbance to the property.

To understand where access is required and what form it might take, it is necessary to

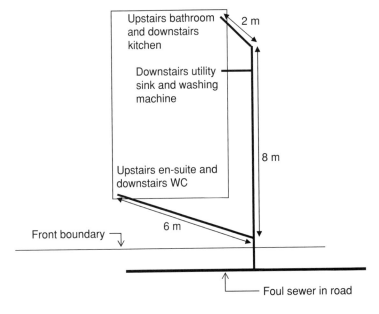

Figure 21.8 Below ground drainage – plan of foul water layout.

Table 21.2 Below ground drainage – access types. Recommendations on dimensions and depths of access types.

Type	Depth to invert	Internal size		Cover size	
		Rectangular	Circular	Rectangular	Circular
Rodding eye	2 m max.	–	100 mm min.	–	100 mm min.
Access fittings					
small[a]	600 mm max.	150 × 100 mm	150 mm	150 × 100 mm	150 mm
large[a]	600 mm max.	225 × 100 mm	150 mm	225 × 100 mm	150 mm
Inspection chamber	600 mm max.	225 × 100 mm	190 mm[b]	225 × 100 mm	190 mm[b]
(shallow)	1 m max.	450 × 450 mm	140 mm	450 × 450 mm	140 mm
Manhole (deep	1.5 m max.	1200 × 750 mm	1050 mm	600 × 600 mm	600 mm
inspection	Over 1.5 m	1200 × 750 mm	1200 mm	600 × 600 mm	600 mm
chamber)	Over 2.7 m[c]	1200 × 840 mm	1200 mm	600 × 600 mm	600 mm

[a] Large access gives greater maximum distance to next access point.
[b] For drains up to 150 mm diameter.
[c] If used with shaft 900 × 840 mm rectangular or 900 mm circular.

know the processes available to clear a blockage in a drain. Traditionally the process involved 'rodding' the drain, but there are now other more flexible clearance methods such as water jetting commonly available in the UK. These new methods require less space within the access, particularly where the drain is relatively close to the surface. The four basic types of access are shown in Table 21.2. There are also limits on the distance between access points to ensure no long lengths of drain are installed where a blockage may occur beyond the reach of the drain clearance process, but these are unlikely to be exceeded on a domestic layout.

The system illustrated in Figure 21.8 is shown in Figure 21.9 in three dimensions, giving the type of access related to depths below ground. Assuming a relatively level site, the building outline is shown at ground floor level. The highest point of the drain is at the connection from the kitchen/bathroom stack and this must therefore be the vent pipe being open to atmospheric pressure, normally the soil vent pipe taken out through the roof. It has been assumed that the depth

Chamber ref.	Invert level	Ground level	Depth
IC1	99.150	99.800	0.600
IC2	98.950	99.750	0.800

Figure 21.9 Foul layout water system with access.

of the sewer has given the final point of access on the property an invert level (level of bottom of the channel formed by the pipe) of 98.950 metres and that this will fix the levels of the drains back through the system. Laying the 100 mm pipes at 1:40 gives inverts and depths below ground as shown in Figure 21.9 on the drawing and in the associated access schedule.

These depths below ground have determined the choice of the type of access required. The utility room connection can be an access fitting with a direct connection to the drain. The soil vent connection has been chosen as inspection chamber 1 (IC1) which, being only 600 mm below ground, could be smaller than inspection chamber 2 (IC2) which is 800 mm below ground, as indicated in Figure 21.9. If the foundations were at 1000 mm below ground all these connections can come out through the wall above the foundations. Since all the connections will be 100 mm pipes, the hole in the wall will have to be formed with lintels with a gap above the pipe to the lintel, so that foundation settlement does not put pressure on the pipes.

This layout with these depths below ground will provide an economic solution with limited excavation and relatively inexpensive access provisions in that deep inspection chambers (manholes) are not necessary. The system is, however, sufficiently far below ground not to be damaged or to be subjected to excessive pressure on the pipe from above ground loadings as long as the pipe and bedding is chosen carefully, as discussed in the next section.

The level of the sewer in the road established the invert levels of the drains in the example. If this sewer had been deeper it would lower the levels throughout the system, making the depth below ground greater, so the access would have to be reconsidered and excavations would be deeper. To keep the system on the property economic it may be better to make the last access (inspection chamber 2 in the example) a backdrop manhole. This keeps the levels and access arrangements the same up to the backdrop where a vertical pipe takes the sewage to a lower level to make the final drain connection to the sewer at a reasonable fall.

Below ground pipes and pipelines

The two main materials used for drains in domestic systems are clay and plastic. Clay gives a rigid pipe that will break with a brittle failure, relying on a crushing strength to avoid damage. The pipes are made in short lengths up to 1.5 metres. Plastic (uPVC) makes a flexible pipe that deforms under load or movement without fracture, so the design needs to ensure that this is limited for the joint to remain watertight. These can be manufactured in long lengths (and are light enough to handle) of up to 6.0 metres.

Both rigid and flexible pipes make flexible pipelines when using jointing systems based on 'O' rings that allow some angular movement without leaking. This will allow the pipeline to move with small settlement and other movements in the ground. The traditional sand and cement joint used on clay pipes will make the whole pipeline rigid, with the most common failure being at the joint. While these sand and cement joints were commonly used in the past they are not now usually specified. Both clay and plastic pipes can be obtained with spigot and socket ends or as plain-ended pipes with a sleeve to make the joint.

This distinction between rigid and flexible pipes is important in deciding on the bedding and the depth of pipes in order to maintain support and protection. The pipe needs to be laid on a stable and accurate surface to lay the pipe to a regular fall and provide sufficient support. The pipe may be vulnerable to uneven loading unless a selected material evenly surrounds the pipe and fully fills the trench in which the pipe is laid. This has given rise to a range of bedding classes offering different levels of support for different soil types and depths as well as depending on the type of pipe. Some of these options for bedding are shown in Figure 21.10.

Rigid pipes with a nominal 100 mm diameter can be laid to within 400 mm of the surface to the top of the pipe using any of the bedding options, although care has to be taken with construction traffic and compaction of trench backfill this close to the surface. If pipes have to be laid very deep (greater than around 8 metres) then the

300 mm

Sidefill →

← All selected as dug material, no stones greater than 40 mm

Trench bottom capable of being trimmed and suitable as dug material available (Class D)

300 mm

Graded all in → granular material

← Selected as dug as above

Bedding 50 mm for sleeve coupling, 100 mm for socket pipes

As dug material available, but poor trimming to trench bottom (Class N)

Class S bed, sidefill and surround

Class B bed and sidefill

Class F bed only

← All single sized granular material

For sites where ground conditions and depths require higher 'bedding factors'

Paving slab

75 mm above crown

Reinforced concrete bridging slab if below road

Material depending on site conditions

Flexible pipes laid close to ground

Figure 21.10   Trench bedding.

strength of the pipe and the bedding specification become structurally important where bedding with granular material has to be specified.

The 100 mm flexible pipe needs to be at least 600 mm below ground to the top of the pipe for fields and gardens (900 mm under roads). Its performance under load is also determined by the width of the trench. The trench needs to be as narrow as possible, with a minimum of the diameter of the pipe plus 300 mm to ensure adequate sidefill. Deep drains are often laid in a trench with working space cut to within around 100 mm of the top of the pipe, with a narrow trench cut to contain the pipe and its bedding. If flexible pipes are laid within 600 mm of the surface (900 mm under roads) then a concrete slab capping has to be provided at least 75 mm above the top of the pipe, as shown in Figure 21.10.

All pipes can be laid close to the surface (or close to building foundations) where the loading and the chance of damage are high by bedding and surrounding the pipe in concrete. This, however, makes the whole pipeline rigid, independent of the pipe material or jointing system.

Life cycle and sustainability

Most of the life cycle issues have already been introduced. The major maintenance task is clearing blockages, so some investment in access can save damage and expense during the operational life. Making the pipeline flexible with the choice of joints between pipes minimises the risk of damage to the joints due to minor movements in the ground, although below ground there is little deterioration in the material as they are designed to carry water.

Drains laid to falls use no energy to move the water, but they do carry the potential for pollution and flooding risks, placing demands on infrastructure works and threatening habitat. Measures to limit the use of water in the house have the greatest influence on the discharge from foul drainage systems, although treatment using septic tanks and/or reed beds would limit the discharge into the public sewer system even further.

Discharge from surface water systems places particular strains on the infrastructure and environment in that they have to be designed to cope with intermittent high flow rates (storms) which increase the risk of flooding from sudden high discharges. Permeable, slow-release systems from hard surfaces, known as sustainable urban drainage systems (SUDS), used on commercial building, particularly on car parks, have application on large-scale housing developments.

Electrical systems

The electrical services are enabling services providing the power to run equipment and appliances to provide the user with both environmental and operational support. Directly involved in lighting and in supplying power to heaters and the central heating boiler and pumps, electricity provides an environmental service. Through power circuits it provides a convenient supply to household equipment such as kitchen and entertainment appliances which satisfies operational needs.

These various uses have different power demands and patterns of use, so it becomes sensible to provide a number of circuits in the house to serve these different purposes:

- Lighting has relatively low power demands, but the majority of lights may be on at the same time (normally taken as 66% of total demand).
- Power for general equipment may include some appliances with high power ratings such as fan heaters and low power ratings such as home entertainment systems. The exact pattern of simultaneous use of these different appliances will be highly variable, but there will be a limit to the number of pieces of equipment being used at any one time for any given floor area in a house.
- Power for single highly rated equipment such as cookers, hot water immersion heaters or instant hot water heaters such as shower units. These require their own circuits.
- Power for units where the pattern of demand is for electricity at night, when demand is nationally low, and it is possible to take advantage of off-peak lower-cost tariffs. This can be used for storage heaters where separate metering arrangements may require separate circuits.

Electricity is dangerous to life. This is not only from direct contact but also from the potential for fire caused by defective systems. It can also damage the appliance or equipment itself. To minimise these risks, circuits have to be provided with the following:

- Earthing to provide an alternative route for the current in the case of individuals or other conducting objects coming into contact with a live circuit. This will lower the risk of serious injury and fire from such contacts.
- Fusing to provide a weak point in the circuit that will fail if a circuit carries too much current by either too high a demand or a fault. This will lower the risk of damage to equipment with the subsequent chance of fire.
- Insulation resistance around wires and at connections to ensure that any contact with the system's components will not result in electric shock or short-circuit.
- Correct polarity to ensure that any switching is provided on the live wire and not the neutral so equipment does not remain live when switched off.
- Sufficient cross-sectional area of the conducting wires to carry the current without overheating causing premature breakdown of insulation or fire.

Supply to the house

The electricity supply for a house will be a single phase providing nominally 230 volts across a pair of wires, the live and the neutral, with a third wire, the earth, to provide continuity back to earth through the supply system. The arrangement for supply and the separation into the required fused circuits is shown in Figure 21.11. The supply company protects its distribution system from excessive demand from a number of houses by installing in each house a fuse that will break the supply if the household takes too much power from the supply at any one time.

Figure 21.11   Electrical systems – supply and circuit distribution.

This is housed in a sealed unit known as the service head, normally provided with an 80- or 100 amp fuse. After the service head, the supply company installs its meter(s) connected to the live and the neutral, and then two wires known as the meter tails are taken from the meter to the consumer unit. The earth is taken to a main earthing terminal.

The earth wiring is kept separate from the live and neutral at this stage to allow earth connections from all potential sources of electric shock. It is possible to get an electric shock not just from direct contact with the live conductor wires but also from indirect conducting systems that might become live from contact with the live wires. The casing around electrical appliances can become live but so can services systems such as the water, heating and gas supply copper piping. The earth wire in the cable will be connected to the casings, giving protection to appliances, but bonding will be required to the copper pipe systems (with appropriate cross bonding between the pipes to sanitary appliances in wet areas). All these connections are made at the main earthing terminal. The direct earth wire is taken to the consumer unit to provide a connection for all the earth wiring in the circuits. Bonding wires are then taken to all potential conducting service systems.

The consumer unit provides a double pole (both live and neutral) isolation switch which is usually a residual current device (RCD) or earth leakage circuit breaker (ELCB), which are sensitive to any earth current and will quickly switch the whole system off automatically if even a small earth current is detected. After the main isolation switch, provision is made to create separate circuits, each with its own over-current protective device (fuse) on the live wire. Originally a rewirable cartridge, the fuse will now most probably be a miniature circuit breaker (MCB). The size of the fuse is determined by the maximum current the circuit is designed to carry.

The power drawn from a circuit by equipment and appliances is rated in watts. Given that the supply is provided nominally at 230 volts, a 1000 watt (1 kilowatt) appliance will create a flow of current of around 4.35 amps (watts = volts × amps). Not all appliances will be used at once, so circuits can be designed on the basis of diversity factors that give domestic circuits fuse sizes as follows:

- Lighting circuits (one for each storey of the house): 5 amps
- Power for general equipment, provided via a ring main serving no more than 100 m² of

floor area: 30 amps (radials are possible, see later in text)

- Power for individual appliances (normal or off-peak tariff) depending on rating of appliance: e.g. cooker 30 or 45 amps, hot water immersion heater 15 amps

A cartridge fuse in the plug should then protect each item of general equipment connected to ring main circuits, which in the UK are rated at up to 13 amps, although 5 and 3 amps should be used for lower rated equipment. Arrangements for both fusing and switching of the larger rated individual appliances (cookers, immersion heaters, etc.) will vary depending on their actual rating, position and with risks associated with wet areas.

Lighting circuits

Figure 21.12 shows a simple lighting circuit with no two-way switching and with each light having an independent switch. The circuit is a radial circuit with the wires taken to connections at each light from the consumer unit out to the last light on the circuit. The live is not taken directly to the lamp at each light connection but via a switch to control each light independently. The earth is also taken to the switch and connected to the switch box to ensure earth continuity if the switch box should become live. These connections at each light are made to a connection strip on the ceiling rose that is mounted to the ceiling at the light fitting. There are enough connections in

the ceiling rose to accommodate two-way switching and multiple lamps on a single switch. An alternative to making the connections in the ceiling rose that is potentially more economic on cable but makes access to the connections more difficult is to use junction boxes (not illustrated). Lighting circuits are on 5 amp fuses from the consumer unit with wires of 1.0 mm^2 for PVC-insulated cable.

Ring mains and radial circuits

Figure 21.13 shows a ring main with, as the name implies, wiring returning to the consumer unit from the last socket outlet. This means that, unlike the lighting circuit, the first wire does not have to carry the full current of all the appliances and equipment on the circuit. Able to flow in both directions, the power will balance itself out in the circuit depending on what equipment is on each socket at any one time. The sizing of the fuse (30 amps) and the wires (2.5 mm^2 PVC-insulated cable) is not dependent on the number of sockets but on the probability of the current from equipment associated with a maximum of 100 m^2 of floor area. Any number of outlets may be installed to ensure the convenience of an outlet for each appliance (or group of equipment as occurs with computer systems) and limit trailing wires. It is possible to gain some economy in the wiring by providing a spur from any of the sockets (or a junction box), but there should only be one spur for each socket outlet on the ring main. Socket outlets may be single or

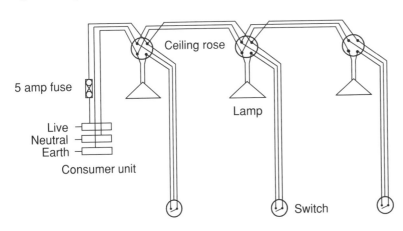

Figure 21.12 Radial lighting circuit.

Figure 21.13 Ring power circuit.

double, will normally be fitted with a switch and are designed to receive a fused plug from the appliance (maximum fuse rating 13 amps). Some fixed equipment may be required to be wired in directly without a plug but still needs to be fused and switched. These are known as fused spur box outlets.

The ring main is economical where sockets are widely distributed as the wiring can be arranged with the first and last socket near the consumer unit. However, radial circuits, where there are no return wires, similar to the lighting circuit, can be used if economies can be made in the wiring. Radial power circuits can be installed with an unlimited number of sockets using 2.5 mm^2 PVC-insulated cable, but can only serve 20 m^2 of floor area and need only be protected with a 20 amp fuse. An alternative to serve a larger floor area up to 50 m^2 with a radial circuit is possible but needs to be protected with a 30 amp fuse and should be wired with 4 mm^2 PVC-insulated cable.

Cables and routing

Decisions on cables and routing are mainly influenced by safety considerations, although maintenance and cost are also factors in making

the choice. The cables are insulated to protect users from shock and short circuit, colour coded to ensure correct polarity and with wires with sufficient cross-sectional area to carry the current in the circuit without overheating when protected by the correct fuse. For domestic installations (lighting and power) cables would normally have PVC insulation around solid copper conductors. The live and the neutral will have individual colour-coded insulation. The colours have been subject to change and it is vital not only to understand the current colour conventions when wiring a new property but also to know past colour coding when working on existing installations.

Maintenance on electrical systems is normally associated with the consumer unit and outlets as this is where all the fusing and connections are made. Although power outlets are usually easily accessible, the consumer unit is normally located high up, out of reach, for safety reasons. Cables are then probably concealed running horizontally in floor and roof spaces. Vertical cabling is then normally buried in a chase in masonry and covered with plaster or placed in the cavity in plasterboarded partitions. Any surface-mounted cables are usually placed in ducts to maintain a tidy appearance. Once cables are hidden then some mechanical protection against accidental damage has to be considered, but, other than the protection of the ducting for surface-mounted cables, this is not normally provided in domestic installations.

Another consideration when concealing cables in floor and roof voids is their positioning with respect to insulation material. If cables are buried in or held up against solid construction by insulation their temperature may rise when carrying current. It may be necessary to increase the size of the conductor wires in the cable in order to limit any potential temperature rise.

Life cycle and sustainability

Considerations of safety and concealment tend to place the components in dry and dark environments, keeping deterioration to a minimum. The design features that limit the risk of fire will minimise the chance of overheating and damaging components with electrical faults. Rewiring may be necessary if insulation efficacy reduces. This is most likely in the flex associated with the drop to the light fitting as this is not only subject to heat, if conventional tungsten bulbs are used, but also to daylight, embrittling the PVC.

Extending the system with additional outlets is not a problem so long as the floor area served by the circuits is not exceeded. The ability to provide new circuits will be limited by the capacity of the consumer unit. Most installations are provided with a consumer unit with spare fused circuit positions.

Although a major system for the distribution of energy, the circuits discussed so far are not consumers of energy, other than the losses in the cables which can be minimised by limiting cable lengths. This will also have an impact on cost and materials used (PVC and copper).

The major contributions to sustainability are in the possibilities for the generation of the electricity from renewable sources and the efficiency of the appliances and equipment specified. This efficiency is not only in the design of the equipment but also in the control over the pattern of use. Installing low energy bulbs in lighting is one step to sustainability, but switching them off when not required may have more impact on overall energy demands. In commercial installations, automatic detection of the need for lighting with movement and light-level sensors that switch lights on and off makes a considerable impact on energy use often because the lighting is required during the day as well as the night. The energy consumption from lighting in domestic systems is less, because daytime lighting is generally not required. Energy saving relies on the occupiers switching off lights when they leave a room. Here education of the user makes the contribution to sustainability. This education of the users has to extend to buying appliances with low energy (and other environmental) ratings. If the appliances are specified with the original building, low energy use should be one of the factors associated with choice.

Having minimised energy and other environmental impacts in the choice of equipment and controls, it is possible to consider generation for an individual property.

Electricity can be generated using photovoltaic arrays, preferably south facing. For an urban property this is most likely to be on the roof, as other positions on the site may be limited. These units generate direct current and therefore an inverter is necessary to provide the AC supply required for the household appliances and equipment.

Photovoltaic units will generate in cloudy conditions albeit at considerably reduced output and in cold weather (where their efficiency is slightly increased), but they cannot generate at night. Their pattern of supply will not match the household demand and therefore a mains connection is still required. However, there will be times during the day when they are generating more electricity than is needed by the household and this over-generation can be sold back to the electricity company. This is achieved through the use of a net meter, where the meter runs backwards when electricity flows back into the supply grid. It is possible to have battery backup to cover loss of supply as, if a net meter is used, the voltaic system has to be shut down, with loss of supply to protect from power surges. To increase the sustainability credentials of photovoltaic units it is claimed that they are made predominantly from silica in a form that is non-polluting and that in good generating conditions (probably not possible in the UK) they can generate the equivalent of their embodied energy in 4 or 5 years and will last for at least 30 years.

Electricity can also be generated from wind energy as another renewable source if the site is suitable. It is also possible to look at waste energy to generate electricity in the use of combined heat and power (CHP) systems. These have been mainly developed for larger community-based energy supply but are available as domestic units.

Summary

(1) Services are active technologies involving movement through distributions systems that can be controlled. The distribution has to be routed through the building, often concealed, but requiring access for maintenance, repair and the renewal of defective parts.

(2) The hot and cold water systems have to deliver water to a variety of appliances at the required quality and at a sufficient rate of flow (quantity) and this will determine most of the basic design of the system. The need for isolation for maintenance and repair then has to be considered. New systems to support sustainable development include low-usage appliances and grey water systems.

(3) The usual expectation of the householder in the case study is for central heating. In the current fuel supply economy this is likely to be gas-fired, pumped hot water to heat radiators in each room. Insulation provides the most effective energy saving investment for a house, although appliance efficiency and controls can make a significant contribution.

(4) The disposal of both foul and surface water is important to health and maintaining a dry environment. Both above and below ground drainage systems are based on moving the water by gravity, by providing falls on the pipes. This restricts the routing and positioning of pipes in order to maintain self-cleansing flow and requires particular attention to be paid to access, to clear any blockage that may occur.

(5) Electrical installations provide power to enable other equipment and appliances to operate. As one of the most dangerous of the service systems, due to its potential to electrocute and start fires, safety determines much of the detail and specification. While equipment and controls can reduce energy consumption, it is possible to consider local generation from renewable resources such as photovoltaic arrays.

22 Guide to Further Reading

The previous chapters described a framework for the analysis of a proposed construction solution in order to make a final choice. The text has introduced many ideas and provided some information on how we build. To make use of this approach in practice, and to keep up to date, you will need to make reference to the wide range of published information. This final chapter gives many such sources and suggests how you might evaluate them to help your overall understanding of construction and make choices for specific projects in the future.

What this book provides

It has been the aim of this book to introduce a comprehensive framework that will allow you to identify the nature and range of knowledge that is necessary to choose construction solutions. The framework suggests that you will require knowledge of what the solution might look like and then the knowledge to undertake an analysis of how it will behave in operation and how it might be constructed.

It is hoped that if you understand what is involved in making a choice you will not only have some idea of what you need to know to make an informed selection but also have a way of integrating new knowledge as you increase your understanding of construction and gain experience.

In exploring the framework and the process of choice, many ideas and concepts, as well as some details and specifications, have been introduced. This represents only a small part of the knowledge required to operate in practice. While your understanding will grow as you gain experience there is still much published material that you will need to make use of.

Sources of information – authority, validity and relevance

You will need to make reference to a wide range of published information for a variety of reasons. You will need to develop your understanding of the ideas related to the basis of the analysis in the earlier chapters of this book and this can be achieved initially with the use of textbooks. The later chapters introduce detailing and specifications for a case study of house construction in the UK. This technical detail is best explored with current publications from a wide variety of sources that identify current regulations and recommendations.

Perhaps the best way to start to think about the range of information that you will need is to understand who prepared the information. This will tell you the purpose of the publication and the basis of the information. This will allow you to match the information you seek to the purpose you have for it.

The publication may be prepared not by an individual, but by an organisation. It may be governmental or commercial. There are many such bodies and associations whose aims include collecting, generating and disseminating information for construction. Some, like the Building Research Establishment (BRE), have broad scope, while others, such as the Concrete Society, are limited. Some are funded by 'independent' sources such as government whereas others are funded by 'vested interest', although this makes the information no less valuable. You will need to identify the scope of their activities, the type and range of their publications and make some assessment of their authority and independence.

Before using any publication, two other factors need to be taken into account when evaluating the relevance of information that an individual publication might have to the knowledge you seek. First, when was it published? In many cases the date of publication is significant, particularly in the use of current regulations and the most up-to-date recommendations. Second, where was it published? Building is a worldwide activity, but not all knowledge transfers between countries, particularly in the case of common practice where climate and the available resource base may not make the solution appropriate for another location.

Textbooks and research papers

Textbooks bring together established knowledge about a whole subject area, e.g. materials, structural design, building services, economics, and will help you understand more about the theory and practice of some of the ideas introduced in this text, particularly in Part 1. You may also need to make reference to other textbooks on construction technology where a greater range of solutions will be introduced and help you understand the development of solutions covered in Part 2. However, the solutions presented in textbooks may have been current when the books were published, but there will have been subsequent changes in legislation and new knowledge and innovations in materials and components. The detailing and specifications suggested in all textbooks, including this book, will need to be confirmed from more recent sources as the time lapsed from the date of publication increases.

Textbooks are normally written by academics and researchers who also write research papers. Research papers are written to contribute to debates, often when knowledge is uncertain and contested. Their purpose is not usually to provide direct advice or recommendations for making selections for construction solutions but to help you understand the nature of the problem and the issues that might be involved. They are often written for the education and research communities, but in practice new or innovative design may require reference to current debate (or past original research findings) which need interpreting in a practical context.

Advice and recommendations – information services

Research papers should not be confused with publications by research bodies, which, after the debates and experiments associated with the research itself, have led to more of a consensus making the knowledge more certain and which will produce publications with the purpose of advice and recommendations for use in practice. Some of the knowledge on which these publications are based will come from experience and studies of observations of past building behaviour where the recommendations have been developed from practice. Such research bodies include:

- The Building Research Establishment (BRE)
- The Construction Industry Research and Information Association (CIRIA)

An example of this would be the BRE *Digests* that provide a concise review of individual subjects from mortars to building on shrinkable clay. The BRE provides a wide range of information directly related to the choice of construction, details and specifications. CIRIA generally provides more comprehensive reports.

Another major group of publications, designed to provide direct advice and recommendations based on research and practice, is provided by organisations concerned primarily with one material and the components that can be created from such materials. They publish information and may provide an advice service to designers for specific projects. The organisations that provide this information service include:

- The British Cement Association (BCA)
- The Concrete Society
- The Steel Construction Institute
- The Timber Research and Development Association (TRADA)
- The Brick Development Association (BDA)
- The Lead Sheet Association
- The Clay Pipe Development Association

Many of these associations, while not providing information about individual manufacturer's products, do represent the interests of commercial organisations in so far as there is a need to provide good information to promote the use of a range of products, if they are to be safely and successfully incorporated into buildings. They bring together technical information and good practice including design data, specification and detailing.

While these organisations are specific to a particular material and their components, they often bring together practice for a particular application. For example, the major publications for timber frame construction are provided by TRADA, but the BDA offers advice and recommendations for cladding a timber frame house in brickwork.

Some of these publications can be based directly on practice, with reports on real projects to provide precedent and examples of how real designs have responded to the brief and the site, giving details and sizes of components that will help in making initial suggestions for similar projects.

Yet another group of organisations that provide an information service are the professional bodies. While some of this service will be exclusive to members, many publish information including technical advice on subjects of interest to their membership. These institutions include all the major construction industry professional bodies.

An example of this type of publication worthy of special mention is the design guide published by the Chartered Institute of Building Services Engineers (CIBSE), which provides not only an explanation of the basis and research behind design procedures for building services but also the data to use in design calculations.

Another handbook of this type but not specific to building services is the *New Metric Handbook*.

Advice and recommendations – manufacturers' information

Many manufacturers provide information about their products. This is not only on the range of products, their shape and size, but also technical information on properties of the materials they use and any manufacturing deviations. They will also provide fixing and jointing requirements.

This should not be confused with their advertising or brochures. These are the technical design manuals giving design data, detailing and even installation and health and safety information.

Many manufacturers include details and suggestions as to how their products can be incorporated into construction to meet performance requirements, particularly where these are set by national regulations. This can be seen in the information provided by both concrete block and insulation manufacturers which gives examples of how to meet current thermal and acoustic regulations.

Manufacturers sometimes provide information on good practice guidance taken from many of the other sources outlined above. Their literature can also often be a good source of references for the appropriate standards and regulations discussed in the next section.

Standards and regulations

It is important to distinguish between regulations made under Act of Parliament, which carry legal obligations in their application and enforcement for all prescribed development, and the standards that may be written into building contracts which are then enforced through that individual contract.

Generally, the regulations made under Act of Parliament are written in performance terms and do not prescribe how the performance will be achieved. However, the major regulations of concern in this text, the Building Regulations, are published with Approved Documents which include construction detail and specifications as well as design information. These suggestions for the construction will satisfy the regulation, but not all situations are covered. Other solutions can be used, but the onus is then on the designer to show performance compliance. This is enforced through the process of building control. The Approved Documents are therefore a good source of information on how construction might meet these performance standards.

For most well-established aspects of construction there will be a British Standard. While these are written to provide a standard that will achieve a performance associated with notional current requirements, their main function is to specify a minimum standard to which a material, component or product should conform. They do not guarantee performance in a specific building. That still has to be judged as part of the final choice. However, these standards do offer advice as to what specification might be applicable to a range of situations and as such provide valuable information on choice as well as a means of specifying for contractual purposes.

Because the development of a full British Standard takes time and often relies on practice and the experience of the committees who draw the standard up, new products may not have a relevant standard to refer to. In this case, British Board of Agrément (BBA) Certificates may be available to outline the nature and use of a particular product.

One set of standards of particular interest to Part 2 of this book are set out in the *National House-Building Council (NHBC) Handbook*. Devised to set standards necessary to gain the warranty cover offered by the NHBC, they bring together many of the recommendations, advice and regulations associated with house construction.

Many of the services installations are covered by specific regulations, an example of which would be the Institute of Electrical Engineers' (IEE) regulations for electrical installations.

Magazines

It is important that the choice of construction is made in the current design and practice context. This is achieved by constantly refreshing the information and knowledge you have about current practice and the social, economic and commercial environments that influence the activity of building. This is possible by working in practice but can be supported by regular reading of a weekly or monthly magazine.

Professional bodies normally provide such a service for their members, but there are a number of commercial magazines that provide not only news but also cover articles on new technical developments and innovative designs

and solutions often associated with particular building projects.

Finding information

It is always possible to seek the information you need by using libraries or directly approaching the organisation or institution that publishes the information. There are, however, databases that allow you to search for and locate information normally based on a common theme or purpose. Originally based on paper, many are now available electronically via the Internet. The electronic storage and access has allowed what where originally indexes, possibly with abstracts outlining the content, to be extended to systems providing full text copies of the information. It is necessary to know who maintains these databases, how they select and authenticate the entries and then make the information available. Many academic and commercially available databases can now be used with confidence, but it is still necessary for you to ensure the authority and relevance of each piece of information for the purpose for which you need it.

These databases should not be confused with 'surfing the net' for information. Using the Internet to access known databases or even the organisations' own information direct from their websites is one thing, but using an open search facility may identify sources of a wide range of authority and relevance. If searches are used, they can identify some excellent information, but much more care is needed to evaluate the material, its source and intention.

Using information – corroboration, confirmation and risk

It may be possible in many cases to find an authoritative and relevant publication and take the information and use it in formulating the solution and then to carry out the analysis to minimise the risk of failure. It is often sufficient to just check the new information against your own existing understanding to confirm it makes sense and then to use it to firm up the solution to be adopted.

However, where this risk is high, or your existing understanding is weak, it may be necessary to seek information from more than one source. It may be that the new information does not seem to make sense with your current understanding so you feel it is necessary to confirm what you are being told.

Asking others for their opinion, if you feel they have the experience and expertise to take a valid view, may satisfy this process of corroboration. However, it might involve seeking other information from published sources.

If you look at a number of sources they may all be quoting one primary source so will not be a fully independent crosscheck. However, if the publications to which you are referring are from authoritative sources, they would have evaluated the validity of the information before they adopted it in their publication. This should give you some confidence in using the information. It is only up to you to confirm it is relevant to your current project.

Remember that it is finally your understanding and good sense that will lead you to make a sound choice.

Summary

(1) This book provides a framework for the analysis and choice of construction. Carrying out the analysis and the process of choice in practice will require reference to a range of current and authoritative publications relevant to the project being considered.

(2) While textbooks and research papers can inform the process of analysis, aspects of detailing and specification have to be current. A range of publications offering current advice and recommendation are available. These are published by organisations that have been established to provide information and by the commercial companies who provide services and products to the industry.

(3) Standards and regulations provide information related to minimum performance achievements that may be used and quoted if they are relevant and appropriate to a specific building project.

(4) Identifying publications is often possible through the use of indexes and libraries, increasingly through electronic sources. Care should be exercised if information is accessed through searches using the Internet rather than established indexes and databases.

(5) It is necessary to make judgements on the quality and relevance of information and it may be necessary to seek corroboration if the risks inherent in the use of the information are high.

Index